I0485834

Artificial Intelligence Applications
Computer Vision Software

Contents

Chapter 1

3DSlicer

3D Slicer (Slicer) is a free and open source software package for image analysis[1] and scientific visualization. Slicer is used in a variety of medical applications, including autism, multiple sclerosis, systemic lupus erythematosus, prostate cancer, schizophrenia, orthopedic biomechanics, COPD, cardiovascular disease and neurosurgery.

1.1   About Slicer

3D Slicer is a free open source software (BSD-style license) that is a flexible, modular platform for image analysis and visualization. 3D Slicer can be extended to enable development of both interactive and batch processing tools for a variety of applications.

3D Slicer provides image registration, processing of DTI (diffusion tractography), an interface to external devices for image guidance support, and GPU-enabled volume rendering, among other capabilities. 3D Slicer has a modular organization that allows the addition of new functionality and provides a number of generic features not available in competing tools.

The interactive visualization capabilities of 3D Slicer include the ability to display arbitrarily oriented image slices, build surface models from image labels, and hardware accelerated volume rendering. 3D Slicer also supports a rich set of annotation features (fiducials and measurement widgets, customized colormaps).

Slicer's capabilities include:[2]

- Handling DICOM images and reading/writing a variety of other formats
- Interactive visualization of volumetric Voxel images, polygonal meshes, and volume renderings
- Manual editing
- Fusion and co-registering of data using rigid and non-rigid algorithms
- Automatic image segmentation
- Analysis and visualization of diffusion tensor imaging data
- Tracking of devices for image-guided procedures.

Slicer is compiled for use on multiple computing platforms, including Windows, Linux, and Mac OS X.

Slicer is distributed under a BSD style, free, open source license. The license has no restrictions on use of the software in academic or commercial projects. However, no claims are made on the software being useful for any particular task. It is entirely the responsibility of the user to ensure compliance with local rules and regulations. Slicer has not been formally approved for clinical use in the FDA in the US or by any other regulatory body elsewhere.

1.2   Image gallery

- Hardware accelerated volume rendering with nVidia drivers, (on Windows and Linux only).

- ProstateNav Module for MRI guided robot assisted biopsy of the prostate.

- Left: 3D rendering. Right: Open MR system

- Visualization of some atlas-based ROIs which correspond to major anatomical fiber tracts. The atlas was provided as part of a download of DTI studio.

- High resolution data acquired on 3-Tesla magnet and post-processed using automated tracking procedure.

- High-dimensional white matter atlas generation and group analysis: result of automatic segmentation of novel subjects.

- Patient-specific modeling in a patient with congenital heart disease.

- Left: Three-dimensional model of levator ani subdivisions including the pubic bone and pelvic viscera. Right: The same model without the pubic bone.

- Cortical parcellations derived from SPGR images obtained from a tumor patient.

- Intraoperative colocalization using iMRI images and 3-D Slicer software.

1.3   History

Slicer started as a masters thesis project between the Surgical Planning Laboratory at the Brigham and Women's Hospital and the MIT Artificial Intelligence Laboratory in 1998.[3] 3D Slicer version 2 has been downloaded several thousand times. In 2007 a completely revamped version 3 of Slicer was released. The next major refactoring of Slicer was initiated in 2009, which aims to transition the GUI of Slicer from using KWWidgets to Qt. Qt-enabled Slicer version 4 has been released in 2011.[4]

Slicer software has enabled a variety of research publications, all aimed at improving image analysis.[5]

This significant software project has been enabled by the participation of several large-scale NIH funded efforts, including the NA-MIC, NAC, BIRN, CIMIT, Harvard Catalyst and NCIGT communities. The funding support comes from several federal funding sources, including NCRR, NIBIB, NIH Roadmap, NCI, NSF and the DOD.

1.4   Users

Slicer's platform provides functionalities for segmentation, registration and three-dimensional visualization of multimodal image data, as well as advanced image analysis algorithms for diffusion tensor imaging, functional magnetic resonance imaging and image-guided radiation therapy. Standard image file formats are supported, and the application integrates interface capabilities to biomedical research software.

Slicer has been used in a variety of clinical research. In image-guided therapy research, Slicer is frequently used to construct and visualize collections of MRI data that are available pre- and intra-operatively to allow for the acquiring of spatial coordinates for instrument tracking.[6] In fact, Slicer has already played such a pivotal role in image-guided therapy, it can be considered as growing up alongside that field, with over 200 publications referencing Slicer since 1998.[7]

In addition to producing 3D models from conventional MRI images, Slicer has also been used to present information derived from fMRI (using MRI to assess blood flow in the brain related to neural or spinal cord activity),[8] DTI (using MRI to measure the restricted diffusion of water in imaged tissue),[9] and electrocardiography.[10] For example, Slicer's DTI package allows the conversion and analysis of DTI images. The results of such analysis can be integrated with the results from analysis of morphologic MRI, MR angiograms and fMRI. Other uses of Slicer include paleontology[11] and neurosurgery planning.[12]

1.5 Developers

The Slicer Developer Orientation offers resources for developers new to the platform. Slicer development is coordinated on the slicer-devel mailing list, and a summary of development statistics is available on Ohloh.

3D Slicer is built on VTK, a pipeline-based graphical library that is widely used in scientific visualization. In version 4, the core application is implemented in C++, and the API is available through a Python wrapper to facilitate rapid, iterative development and visualization in the included Python console. The user interface is implemented in Qt, and may be extended using either C++ or Python.

Slicer supports several types of modular development. Fully interactive, custom interfaces may be written in C++ or Python. Command-line programs in any language may be wrapped using a light-weight XML specification, from which a graphical interface is automatically generated.

For modules that are not distributed in the Slicer core application, a system is available to automatically build and distribute for selective download from within Slicer. This mechanism facilitates the incorporation of code with different license requirements from the permissive BSD-style license used for the Slicer core.

The Slicer build process utilizes CMake to automatically build prerequisite and optional libraries (excluding Qt). The core development cycle incorporates automatic testing, as well as incremental and nightly builds on all platforms, monitored using an online dashboard.

1.6 Criticism

Development still in progress, Slicer is sometimes accused by users of being poorly documented and a lacking in automation facilities (which is useful in batch processing). Other users report that Slicer has excellent documentation and training materials. Also Slicer's user interface and internal processing logic is fully scriptable. Although bugs can be reported to the mailing list and issue tracker, they are addressed based on developer availability. Updated versions are periodically released with updated features, while the development version with the latest source code is available daily.

1.7 External dependencies

- VTK

- ITK

- CMake

- CPack

- Python

- Tcl

- Nrrd

- MRML (MRML)

- IGSTK

- KWWidgets

- Qt

1.8   See also

- Analyze

- GIMIAS

- Mimics

1.9   References

[1] Pieper S., Halle M., Kikinis R. 3D SLICER. Proceedings of the 1st IEEE International Symposium on Biomedical Imaging: From Nano to Macro 2004; 1:632–635.

[2] Pieper S., Lorensen B., Schroeder W., Kikinis R. The NA-MIC Kit: ITK, VTK, Pipelines, Grids and 3D Slicer as an Open Platform for the Medical Image Computing Community. Proceedings of the 3rd IEEE International Symposium on Biomedical Imaging: From Nano to Macro 2006; 1:698-701.

[3] Hirayasu, Y; Shenton, ME; Salisbury, DF; Dickey, CC; Fischer, IA; Mazzoni, P; Kisler, T; Arakaki, H; Kwon, JS; Anderson, JE; Yurgelun-Todd, D; Tohen, M; McCarley, RW (1998). "Lower left temporal lobe MRI volumes in patients with first-episode schizophrenia compared with psychotic patients with first-episode affective disorder and normal subjects". *The American Journal of Psychiatry* **155** (10): 1384–91. PMID 9766770.

[4] Fedorov; Beichel; Kalpathy-Cramer; Finet; Fillion-Robin; Pujol; Bauer; Jennings; Fennessy; Sonka; Buatti; Aylward; Miller; Pieper; Kikinis (2012). "3D Slicer as an image computing platform for the Quantitative Imaging Network". *Magnetic Resonance Imaging* **30** (9): 1323–41. doi:10.1016/j.mri.2012.05.001. PMID 22770690.

[5] Pieper S., Lorensen B., Schroeder W., Kikinis R. The NA-MIC Kit: ITK, VTK, Pipelines, Grids and 3D Slicer as an Open Platform for the Medical Image Computing Community. Proceedings of the 3rd IEEE International Symposium on Biomedical Imaging: From Nano to Macro 2006; 1:698–701.

[6] Hata, N; Piper, S; Jolesz, FA; Tempany, CM; Black, PM; Morikawa, S; Iseki, H; Hashizume, M; Kikinis, R (2007). "Application of open source image guided therapy software in MR-guided therapies". *Medical image computing and computer-assisted intervention : MICCAI ... International Conference on Medical Image Computing and Computer-Assisted Intervention* **10** (Pt 1): 491–8. doi:10.1007/978-3-540-75757-3_60. PMID 18051095.

[7] For a list of publications citing Slicer usage since 1998, visit: http://www.slicer.org/publications/pages/display/?collectionid=11

[8] Archip, N; Clatz, O; Whalen, S; Kacher, D; Fedorov, A; Kot, A; Chrisochoides, N; Jolesz, F; Golby, A; Black, PM; Warfield, SK (2007). "Non-rigid alignment of pre-operative MRI, fMRI, and DT-MRI with intra-operative MRI for enhanced visualization and navigation in image-guided neurosurgery". *NeuroImage* **35** (2): 609–24. doi:10.1016/j.neuroimage.2006.11.060. PMC 3358788. PMID 17289403.

[9] Ziyan, U; Tuch, D; Westin, CF (2006). "Segmentation of thalamic nuclei from DTI using spectral clustering". *Medical image computing and computer-assisted intervention : MICCAI ... International Conference on Medical Image Computing and Computer-Assisted Intervention* **9** (Pt 2): 807–14. doi:10.1007/11866763_99. PMID 17354847.

[10] Verhey, JF; Nathan, NS; Rienhoff, O; Kikinis, R; Rakebrandt, F; D'ambra, MN (2006). "Finite-element-method (FEM) model generation of time-resolved 3D echocardiographic geometry data for mitral-valve volumetry". *Biomedical engineering online* **5**: 17. doi:10.1186/1475-925X-5-17. PMC 1421418. PMID 16512925.

[11] http://openpaleo.blogspot.com/2009/03/3d-slicer-tutorial-part-vi.html

[12] http://picasaweb.google.com/107065747472066371420

1.10   External links

- Slicer

Chapter 2

AForge.NET

AForge.NET is a computer vision and artificial intelligence library originally developed by Andrew Kirillov for the .NET Framework.

The source code and binaries of the project are available under the terms of the Lesser GPL and the GPL (GNU General Public License).

Another (unaffiliated) project named **Accord.NET** extends the features of the original **AForge.NET** library.

2.1 Features

The framework's API includes support for:

- Computer vision, image processing and video processing

    - Including a comprehensive image filter library

- Neural networks

- Genetic programming

- Fuzzy logic

- Machine learning

- and libraries for a select set of robotics kits

    - Lego Mindstorms NXT and RCX kits
    - TeRK Qwerk kit
    - Surveyor SRV-1 and SVS kits

Complete list of features is available on the features page of the project.

The framework is provided not only with different libraries and their sources, but with many sample applications, which demonstrate the use of this framework, and with documentation help files, which are provided in HTML Help format. The documentation is also available on-line.

2.2   See also

- OpenCV - A popular C++ computer vision library.

- VXL - Another C++ computer vision library.

- CVIPtools - A complete GUI based computer vision and image processing software environment.

- OpenNN - An open source C++ neural networks library.

2.3   References

[1] End of free public support, April 1, 2012, AForge.NET

[2] http://aforgenet.com/news/2011.12.21.five_years_framework.html

2.4   External links

- Official website

- Google Code project home

- Accord.NET Website

Chapter 3

Amira (software)

Amira (pronounce: Ah-meer-ah) is a software platform for 3D and 4D data visualization, processing, and analysis. It is being actively developed by Visualization Sciences Group, Bordeaux, France and the Zuse Institute Berlin (ZIB), Germany.

3.1 Overview

Amira[1] is an extendable software system for scientific visualization, data analysis, and presentation of 3D and 4D data. Amira is being developed and commercially distributed by FEI Visualization Sciences Group, Bordeaux in cooperation with the Zuse Institute Berlin (ZIB). It is used by several thousand researchers and engineers in academia and industry around the world. Amira's flexible user interface and its modular architecture make it a universal tool for processing and analysis of data from various modalities; e.g. micro-CT,[2] PET,[3] Ultrasound.[4] Its ever expanding functionality has made it a versatile data analysis and visualization solution, applicable to and being used in many fields, such as microscopy in biology[5] and materials science,[6] molecular biology,[7] quantum physics,[8] astrophysics,[9] computational fluid dynamics (CFD),[10] finite element modeling (FEM),[11] non-destructive testing (NDT),[12] and many more. One of the key features, besides data visualization, is Amira's set of tools for image segmentation[13] and geometry reconstruction.[14] This allows the user to mark (or segment) structures and regions of interest in 3D image volumes using automatic, semi-automatic, and manual tools. The segmentation can then be used for a variety of subsequent tasks, such as volumetric analysis,[4] density analysis,[15] shape analysis,[16] or the generation of 3D computer models for visualization,[17] numerical simulations,[18] or rapid prototyping[19] or 3D printing, to name a few. Other key Amira features are multi-planar and volume visualization, image registration,[20] filament tracing,[21] cell separation and analysis,[16] tetrahedral mesh generation,[22] fiber-tracking from diffusion tensor imaging (DTI) data,[23] skeletonization,[24] spatial graph analysis, and stereoscopic rendering[25] of 3D data over multiple displays including CAVEs (Cave automatic virtual environments).[26] As a commercial product Amira requires the purchase of a license or an academic subscription. A time-limited, but full-featured evaluation version is available for download free of charge.

3.2 History

3.2.1 1994–1998 Research Software

Amira's roots go back to 1994 and the Department for Scientific Visualization, headed by Hans-Christian Hege at the Zuse Institute Berlin (ZIB). The ZIB is a research institute for mathematics and informatics. The Scientific Visualization department's mission is to help solve computationally and scientifically challenging tasks in medicine, biology, and engineering. For this purpose, it develops algorithms and software for 2D, 3D, and 4D data visualization and visually supported exploration and analysis. At that time, the young visualization group at the ZIB had experience with the extendable, data flow-oriented visualization environments apE,[27] IRIS Explorer,[28] and Advanced Visualization Studio (AVS), but was not satisfied with these products' interactivity, flexibility, and ease-of-use for non-computer scientists.

Therefore, in a subproject[29] within a medically oriented, multi-disciplinary collaborative research center[30] the development of a new software system was started in early 1994. The initial development was performed by Detlev Stalling, who later became the chief software architect. The software system was called "HyperPlan", highlighting its initial target application – a planning system for hyperthermia cancer treatment. The system was being developed on Silicon Graphics (SGI) computers, which at the time were the standard workstations used for high-end graphics computing. Software development was based on libraries such as OpenGL, SGI Open Inventor, and the graphical user interface libraries X11, Motif (software), and ViewKit. In 1998, X11/Motif/Viewkit were replaced by the Qt toolkit.

The HyperPlan framework served as the base for more and more projects at the ZIB and was used by a growing number of researchers in collaborating institutions. The projects included applications in neurobiology, confocal microscopy, flow visualization, molecule visualization and analysis and computational astrophysics.

3.2.2   1998–today Commercially Supported Product

The growing number of users of the system started to exceed the capacities that ZIB could spare for software distribution and support, as ZIB's primary mission was algorithmic research. Therefore the spin-off company Indeed – Visual Concepts GmbH was founded by Hans-Christian Hege, Detlev Stalling, and Malte Westerhoff with the vision of making the extensive capabilities of the software available to researchers in industry and academia worldwide and to provide the product support and robustness needed in today's fast-paced and competitive world.

In Feb 1998 the HyperPlan software was given the new, less application-specific name "Amira". This name is not an acronym but was chosen for being pronounceable in different languages, starting with an 'A', and having an appropriate connotation: the Latin verb "admirare" (to admire), meaning "to look at" and "to wonder at", describes a typical situation in data visualization.

A major re-design of the software was undertaken by Detlev Stalling and Malte Westerhoff in order to make it a commercially supportable product and to make it available on non-SGI computers as well. In March 1999, the first version of the commercial Amira was shown at the CeBIT tradeshow in Hannover, Germany on SGI IRIX and Hewlett-Packard UniX (HP-UX). Versions for Linux and Microsoft Windows followed within the following twelve months. Later Mac OS X support was added. Indeed – Visual Concepts selected the Bordeaux, France and San Diego, USA based company TGS, Inc. as the worldwide distributor for Amira and completed five major releases (up to version 3.1) in the subsequent four years.

In 2003 both Indeed as well as TGS were acquired by Massachusetts-based Mercury Computer Systems, Inc. (NASDAQ: MRCY) and became part of Mercury's newly formed life sciences business unit, later branded Visage Imaging. In 2009, Mercury Computer Systems, Inc. spun off Visage Imaging again and sold it to Melbourne, Australia based Promedicus Ltd (ASX:PME), a leading provider of radiology information systems and medical IT solutions. During this time, Amira continued to be developed in Berlin, Germany and in close collaboration with the ZIB, still headed by the original creators of Amira. TGS, located in Bordeaux, France was sold by Mercury Computer systems to a French investor and renamed to Visualization Sciences Group (VSG). VSG continued the work on a complementary product named Avizo, based on the same source code but customized for material sciences.

In August 2012, FEI, to that date the largest OEM reseller of Amira, purchased VSG and the Amira business from Promedicus. In August 2013 Visualization Sciences Group (VSG) has been renamed to FEI Visualization Sciences Group. Amira and Avizo are still being marketed as two different products; Amira for life sciences and Avizo for material sciences, but the development efforts are now joined once again. As in the beginning, the Amira roadmap continues to be driven by the interesting and challenging scientific questions that Amira users around the world are trying to answer, often at the leading edge in their fields.

The latest version, Amira 5.4.3, was released in October 2012, with the next release planned for the second quarter of 2013.

3.3   Amira Options

3.3.1 Microscopy Option

- Specific readers for microscopy data

- Image deconvolution

- Exploration of 3D imagery obtained from virtually any microscope

- Extraction and editing of filament networks from microscopy images

3.3.2 DICOM Reader

- Import of clinical and preclinical data in DICOM format

3.3.3 Mesh Option

- Generation of 3D finite element (FE) meshes from segmented image data

- Support for many state-of-the-art FE solver formats

- High-quality visualization of simulation mesh-based results, using scalar, vector, and tensor field display modules

3.3.4 Skeletonization Option

- Reconstruction and analysis of neural and vascular networks

- Visualization of skeletonized networks

- Length and diameter quantification of network segments

- Ordering of segments in a tree graph

- Skeletonization of very large image stacks

3.3.5 Molecular Option

- Advanced tools for the visualization of molecule models

- Hardware-accelerated volume rendering

- Powerful molecule editor

- Specific tools for complex molecular visualization

3.3.6 Developer Option

- Creation of new custom components for visualizing or data processing

- Implementation of new file readers or writers

- C++ programming language

- Development wizard for getting started quickly

3.3.7   Neuro Option

- Medical image analysis for DTI and brain perfusion

- Fiber tracking supporting several stream-line based algorithms

- Fiber separation into fiber bundles based on user defined source and destination regions

- Computation of tensor fields, diffusion weighted maps

- Eigenvalue decomposition of tensor fields

- Computation of mean transit time, cerebral blood flow, and cerebral blood volume

3.3.8   VR Option

- Visualization of data on large tiled displays or in immersive Virtual Reality (VR) environments

- Support of 3D navigation devices

- Fast multi-threaded and distributed rendering

3.3.9   Very Large Data Option

- Support for visualization of image data exceeding the available main memory, using efficient out-of-core data management

- Extensions of many standard modules, such as orthogonal and oblique slicing, volume rendering, and isosurface rendering, to work on out-of-core data

3.4   Editors

- **CameraPath Editor:** create a camera path using key-frames for animations and movies

- **Color Dialog:** define a color value using a graphical interface

- **2 Colormap Editors:** modify the RGBA values of a discrete colormap

- **Curve Editor:** create and edit curves

- **Demo Manager:** manage and control demos using a graphical interface

- **Digital Image Filters:** apply standard image processing filters

- **Filament Editor:** skeletonize image data and modify spatial graphs

- **Grid Editor:** edit and simplify tetrahedral grids

- **Image Crop Editor:** crop 3D images, change bounding box, and voxel size

- **Landmark Editor:** add, move, or delete markers in a landmark set

- **LineSet Editor:** select, create, modify, and delete polylines

- **Multi-planar Viewer:** view up to two data sets simultaneously in a 3+1 MPR viewer

- **Parameter Editor:** add, change, or delete attributes of a data object

- **Plot Tool:** display 2D plots

- **Segmentation Editor:** 3D image segmentation using interactive and semi-automatic tools

- **Surface Simplification Editor:** reduce the number of triangles of a triangulated surface

- **Surface Editor:** modify triangles, remove intersections, assign boundary ids in triangulated surfaces

- **Transform Editor:** translate, rotate, or scale any 3D data object

3.5 Application Areas

- Anatomy[31][32]

- Biochemistry[33]

- Biophysics[33]

- Cellular microbiology[34][35]

- Computational fluid dynamics[36]

- Cryo-electron tomography[34]

- Diffusion MRI/Fiber Tracking

- Embryology[31]

- Endocrinology[37]

- Finite Element Modelling[38]

- Histology[31][33][39]

- Medical imaging research

- Microscopy in life and material sciences

- Molecular biology[40]

- Neuroscience[39][41]

- Nondestructive testing

- Orthopedics[38][42][43]

- Otolaryngology[44]

- Preclinical imaging[40]

- Urology[45]

3.6 Processing and Data Analysis

- Surface and grid generation

- 3D image segmentation

- Image registration and slice alignment

- Skeletonization and deconvolution

- Multitude of quantification tools

- Arithmetic operations

- MATLAB integration

- 2D and 3D image filtering

- Surface generation

- Finite element model (FEM) grid generation

- Interactive and automatic segmentation

- Interactive and automatic slice alignment

- Image registration and morphing

- Tensor computation

- Skeletonization and tracing of neural and vascular networks

- Deconvolution and z-drop correction

- Powerful scripting interface

- Dedicated editors for segmentation, tracing, and fusion

3.7   Visualization

- Orthogonal and oblique slicing

- Volume rendering

- Surface rendering

- Isolines and isosurfaces

- Multi-channel imaging and fusion

- Vector and tensor visualization

- Support of structured / unstructured grids

- Molecular simulation and visualization

- Structured workflow visualization

- Active and passive stereo support

- Tiled screen support

- Virtual reality navigation and tools

3.8 Presentation

- Easy-to-use interactive 3D navigation
- Tools for designing animated demos
- Automation of complex animations and demonstrations
- Embedded tools for movie generation
- Active and passive 3D stereo vision
- 2D and 3D annotation
- Support for stereoscopic and auto-stereoscopic displays
- Virtual reality navigation tools
  - Single and tiled screen display
  - Single or multi-pipe rendering
  - Support for "trackd" input devices
- Geometry data
- Scalar fields and all types of multidimensional images
- Vector and flow data
- Tensor fields
- Molecular models
- Simulation data on finite element models

3.9 Supported File Formats

3.10 Release history

3.11 References

[1] Stalling, D.; Westerhoff, M.; Hege, H.-C. (2005). C.D. Hansen and C.R. Johnson, ed. "Amira: A Highly Interactive System for Visual Data Analysis". *The Visualization Handbook* (Elsevier): 749–767. CiteSeerX: 10.1.1.129.6785.

[2] Adam, R.; Smith, A.R.; Sieren, J.C.; Eggleston, T.; McLennan, G. (2010). "Characterization Of The Airways And Lungs For The FABP/CFTR-Knockout Mouse Using Micro-Computed Tomography And Large Image Microscope Array" (PDF). *American Journal of Respiratory and Critical Care Medicine* (Am Thoracic Soc) **181**: A6264. doi:10.1164/ajrccm-conference.2010.181.1_meetingabstracts

[3] Awasthi, V.; Holter, J.; Thorp, K.; Anderson, S.; Epstein, R. (2010). "F-18-fluorothymidine-PET evaluation of bone marrow transplant in a rat model". *Nuclear Medicine Communications* **31** (2): 152. doi:10.1097/mnm.0b013e3283339f92.

[4] Ayers, G.D.; McKinley, E.T.; Zhao, P.; Fritz, J.M.; Metry, R.E.; Deal, B.C.; Adlerz, K.M.; Coffey, R.J.; Manning, H.C. (2010). "Volume of Preclinical Xenograft Tumors Is More Accurately Assessed by Ultrasound Imaging Than Manual Caliper Measurements". *Journal of Ultrasound in Medicine* (Am inst Ultrasound Med) **29** (6): 891.

[5] Dlasková, A.; Spacek, T.; Santorová, J.; Plecitá-Hlavatá, L.; Berková, Z.; Saudek, F.; Lessard, M.; Bewersdorf, J.; Jezek, P. (2010). "4Pi microscopy reveals an impaired three-dimensional mitochondrial network of pancreatic islet beta-cells, an experimental model of type-2 diabetes.". *Biochimica et Biophysica Acta (BBA)-Bioenergetics* (Elsevier). doi:10.1016/j.bbabio.2010.02.003.

[6] Clark, N.D.L.; Daly., C. (2010). "Using confocal laser scanning microscopy to image trichome inclusions in amber" (PDF). *Journal of Paleontological Techniques* **8**.

[7] Amstalden van Hove, E.R.; Blackwell, T.R.; Klinkert, I.; Eijkel, G.B.; Heeren, R.; Glunde, K. (2010). "Multimodal Mass Spectrometric Imaging of Small Molecules Reveals Distinct Spatio-Molecular Signatures in Differentially Metastatic Breast Tumor Models". *Cancer Research* (AACR) **70** (22): 9012. doi:10.1158/0008-5472.can-10-0360.

[8] Sherman, D.M. (2010). "Metal complexation and ion association in hydrothermal fluids: insights from quantum chemistry and molecular dynamics.". *Geofluids* (John Wiley & Sons) **10** (1-2): 41–57. doi:10.1002/9781444394900.ch4.

[9] O'Neill, S.M.; Jones, T.W. (2010). "Three-Dimensional Simulations of Bi-Directed Magnetohydrodynamic Jets Interacting with Cluster Environments.". *The Astrophysical Journal* (IOP Publishing) **710** (1): 180. arXiv:1001.1747. Bibcode:2010ApJ...710..180O. doi:10.1088/0004-637x/710/1/180.

[10] Baharoglu, M.I.; Schirmer, C.M.; Hoit, D.A.; Gao, B.L.; Malek, A.M. (2010). "Aneurysm Inflow-Angle as a Discriminant for Rupture in Sidewall Cerebral Aneurysms". *Morphometric and Computational Fluid Dynamic Analysis* (Stroke, Am Heart Assoc).

[11] Bardyn,, T.; Gédet, P.; Hallermann, W.; Büchler., P. (2010). "Prediction of dental implant torque with a fast and automatic finite element analysis: a pilot study.". *Oral Surgery, Oral Medicine, Oral Pathology, Oral Radiology, and Endodontology* (Elsevier). doi:10.1016/j.tripleo.2009.11.010.

[12] Shearing, P.R.; Gelb, J.; Yi, J.; Lee, W.K.; Drakopolous, M.; Brandon, N.P. (2010). "Analysis of Triple Phase Contact in Ni-YSZ Microstructures Using Non-destructive X-ray Tomography with Synchrotron Radiation". *Electrochemistry Communications* (Elsevier). doi:10.1016/j.elecom.2010.05.014.

[13] Jährling, N.; Becker, K.; Schönbauer, C.; Schnorrer, F.; Dodt, H.U. (2010). "Three-dimensional reconstruction and segmentation of intact Drosophila by ultramicroscopy". *Frontiers in Systems Neuroscience* (Frontiers Research Foundation) **4**: 1. doi:10.3389/neuro.06.001.2010. PMC 2831709. PMID 20204156.

[14] Zheng, G. (2010). "Statistical shape model-based reconstruction of a scaled, patient-specific surface model of the pelvis from a single standard AP x-ray radiograph.". *Medical Physics* **37**: 1424. Bibcode:2010MedPh..37.1424Z. doi:10.1118/1.3327453.

[15] Rodriguez-Soto, A.E.; Fritscher, K.D.; Schuler, B.; Issever, A.S.; Roth, T.; Kamelger, F.; Kammerlander, C.; Blauth, M.; Schubert, R.; Link, T.M. (2010). "Texture Analysis, Bone Mineral Density, and Cortical Thickness of the Proximal Femur: Fracture Risk Prediction.". *Journal of Computer Assisted Tomography* **34** (6): 949. doi:10.1097/rct.0b013e3181ec05e4.

[16] Leischner, U.; Schierloh, A.; Zieglgänsberger, W.; Dodt, H.U. (2010). "Formalin-Induced Fluorescence Reveals Cell Shape and Morphology in Biological Tissue Samples". *Public Library of Science* **5** (4): e10391. Bibcode:2010PLoSO...510391L. doi:10.1371/journal.pone.0010391. PMC 2861007. PMID 20436930.

[17] Felts, R.L.; Narayan, K.; Estes, J.D.; Shi, D.; Trubey, C.M.; Fu, J.; Hartnell, L.M.; Ruthel, G.T.; Schneider, D.K.; Nagashima, K. (2010). "3D visualization of HIV transfer at the virological synapse between dendritic cells and T cells.". *Proceedings of the National Academy of Sciences of the United States of America* (National Acad Sciences) **107** (30): 13336. Bibcode:2010PNAS..10713336F. doi:10.1073/pnas.1003040107.

[18] Taylor, D.J.; Doorly, D.J.; Schroter, R.C. (2010). "Inflow boundary profile prescription for numerical simulation of nasal airflow.". *Journal of the Royal Society Interface* (The Royal Society) **7** (44): 515. doi:10.1098/rsif.2009.0306.

[19] Lucas, B.C.; Bogovic, J.A.; Carass, A.; Bazin, P.L.; Prince, J.L.; Pham, D.L.; Landman, B.A. (2010). "The Java Image Science Toolkit (JIST) for Rapid Prototyping and Publishing of Neuroimaging Software" (PDF). *Neuroinformatics* (Springer) **8** (1): 5–17. doi:10.1007/s12021-009-9061-2.

[20] Dasgupta, S.; Feleppa, E.; Ramachandran, S.; Ketterling, J.; Kalisz, A.; Haker, S.; Tempany, C.; Porter, C.; Lacrampe, M.; Isacson, C. (2007). "8A-4 Spatial Co-Registration of Magnetic Resonance and Ultrasound Images of the Prostate as a Basis for Multi-Modality Tissue-Type Imaging". pp. 641–643.

[21] Oberlaender, M.; Bruno, R.M.; Sakmann, B.; Broser, P.J. (2007). "Transmitted light brightfield mosaic microscopy for three-dimensional tracing of single neuron morphology.". *Journal of Biomedical Optics* **12**: 064029. Bibcode:2007JBO....12f4029O. doi:10.1117/1.2815693.

[22] Lamecker, H.; Mansi, T.; Relan, J.; Billet, F.; Sermesant, M.; Ayache, N.; Delingette., H. (2009). "Adaptive Tetrahedral Meshing for Personalized Cardiac Simulations.". *Citeseer*.

[23] Boretius, S.; Michaelis, T.; Tammer, R.; Ashery-Padan, R.; Frahm, J.; Stoykova, A. (2009). "In vivo MRI of altered brain anatomy and fiber connectivity in adult pax6 deficient mice.". *Cerebral Cortex* (Oxford University Press) **19** (12): 2838. doi:10.1093/cercor/bhp057.

[24] Kohjiya, S.; Katoh, A.; Suda, T.; Shimanuki, J.; Ikeda, Y. (2006). "Visualisation of carbon black networks in rubbery matrix by skeletonisation of 3D-TEM image.". *Polymer* (Elsevier) **47** (10): 3298–3301. doi:10.1016/j.polymer.2006.03.008.

[25] Clements, R.J.; Mintz, E.M.; Blank, J.L. (2009). "High resolution stereoscopic volume visualization of the mouse arginine vasopressin system.". *Journal of neuroscience methods* (Elsevier). doi:10.1016/j.jneumeth.2009.12.011.

[26] Ohno, N.; Kageyama., A. (2009). "Region-of-interest visualization by CAVE VR system with automatic control of level-of-detail.". *Computer Physics Communications* (Elsevier). Bibcode:2010CoPhC.181..720O. doi:10.1016/j.cpc.2009.12.002.

[27] Dyer, D.S. (1990). "A dataflow toolkit for visualization.". *Computer Graphics and Applications,* (IEEE): 60–69. doi:10.1109/38.56300.

[28] Foulser, D. (1995). "IRIS Explorer: A framework for investigation.". *Computer Graphics,* (ACM SIGGRAPH): 13–16. doi:10.1145/204362.204365.

[29] "DFG Project: Algorithmen zur Planung und Kontrolle von Hyperthermiebehandlungen". DFG Deutsche Forschungsgemeinschaft. Retrieved 28 January 2015.

[30] "DFG Project SFB 273: Hyperthermia: Methodics and Clinics". DFG Deutsche Forschungsgemeinschaft. Retrieved 28 January 2015.

[31] de Boer, B.A.; Soufan, A.T.; Hagoort, J.; Mohun, T.J.; van den Hoff, M.J.B; Hasman, A.; Voorbraak, F.P.J.M.; Moorman, A.F.M.; Ruijter, J.M. (2011). "The interactive presentation of 3D information obtained from reconstructed datasets and 3D placement of single histological sections with the 3D portable document format.". *Development* **138** (1): 159. doi:10.1242/dev.051086.

[32] Specht, M.; Lebrun, R.; Zollikofer, C.P.E. (2007). "Visualizing shape transformation between chimpanzee and human braincases." (PDF). *The Visual Computer: International Journal of Computer Graphics archive* **23** (9): 743–751. doi:10.1007/s00371-007-0156-1.

[33] Gaemers, I.C.; Stallen, J.M.; Kunne, C.; Wallner, C.; van Werven, J.; Nederveen, A.; Lamers, W.H. (2011). "Lipotoxicity and steatohepatitis in an overfed mouse model for non-alcoholic fatty liver disease.". *Biochimica et Biophysica Acta (BBA)-Molecular Basis of Disease* (Elsevier). doi:10.1016/j.bbadis.2011.01.003.

[34] Kudryashev, M; Cyrklaff, M.; Alex, B.; Lemgruber, L.; Baumeister, W.; Wallich, R.; Frischknecht, F. (2011). "Evidence of direct cell-cell fusion in Borrelia by cryogenic electron tomography.". *Cellular Microbiology* (Wiley Online Library). doi:10.1111/j.1462-5822.2011.01571.x.

[35] Meisslitzer-Ruppitsch, C.; Röhrl, C.; Ranftler, C.; Neumüller, J.; Vetterlein, M.; Ellinger, A.; Pavelka, M. (2011). "The ceramide-enriched trans-Golgi compartments reorganize together with other parts of the Golgi apparatus in response to ATP-depletion.". *Histochemistry and Cell Biology* (Springer) **135** (2): 159–171. doi:10.1007/s00418-010-0773-z.

[36] Bevan, R.L.T.; Sazonov, I.; Saksono, P.H.; Nithiarasu, P.; van Loon, R.; Luckraz, H.; Ashral, S. (2011). "Patient-specific blood flow simulation through an aneurysmal thoracic aorta with a folded proximal neck.". *Numerical Methods in Biomedical Engineering* (Wiley) **27** (8): 1167–1184. doi:10.1002/cnm.1425.

[37] Bujotzek, A.; Shan, M.; Haag, R.; Weber, M. (2011). "Towards a rational spacer design for bivalent inhibition of estrogen receptor". *Journal of Computer-Aided Molecular Design* **25** (3): 253–262. Bibcode:2011JCAMD..25..253B. doi:10.1007/s10822-011-9417-1.

[38] Cai, W.; Lee, E.Y.; Vij, A.; Mahmood, S.A.; Yoshida, H. (2011). "MDCT for Computerized Volumetry of Pneumothoraces in Pediatric Patients.". *Academic Radiology* (Elsevier).

[39] Irving, S.; Moore, D.R.; Liberman, M.C.; Sumner, C.J. (2011). "Olivocochlear Efferent Control in Sound Localization and Experience-Dependent Learning.". *Journal of Neuroscience* (Soc Neuroscience) **31** (7): 2493. doi:10.1523/jneurosci.2679-10.2011.

[40] Obenaus, A.; Hayes, P. (2011). "Drill hole defects: induction, imaging, and analysis in the rodent.". *Methods in molecular biology* (Springer) **690**: 301. doi:10.1007/978-1-60761-962-8_20.

[41] Ertürk, A.; Mauch, C.P.; Hellal, F.; Förstner, F.; Keck, T.; Becker, K.; Jährling, N.; Steffens, H.; Richter, M.; Hübener, M.; Kramer, E.; Kirchhoff, F.; Dodt; Bradke, F. (2011). "Three-dimensional imaging of the unsectioned adult spinal cord to assess axon regeneration and glial responses after injury.". *Nature Medicine*. doi:10.1038/nm.2600.

[42] Carlson, K.J.; Wrangham, R.W.; Muller, M.N.; Sumner, D.R.; Morbeck, M.E.; Nishida, T.; Yamanaka, A.; Boesch, C. (2011). "Comparisons of Limb Structural Properties in Free-ranging Chimpanzees from Kibale, Gombe, Mahale, and Tai Communities.". *Primate Locomotion* (Springer): 155–182. doi:10.1007/978-1-4419-1420-0_9.

[43] Hartwig, T.; Streitparth, F.; Gro, C.; Müller, M.; Perka, C.; Putzier, M.; Strube, P. (2011). "Digital 3-Dimensional Analysis of the Paravertebral Lumbar Muscles After Circumferential Single-level Fusion.". *Journal of Spinal Disorders & Techniques.*

[44] Lee, J.; Eddington, D.K.; Nadol, J.B. (2011). "The Histopathology of Revision Cochlear Implantation.". *Audiology and Neurotology* **16** (5): 336–346. doi:10.1159/000322307.

[45] Han, M.; Kim, C.; Mozer, P.; Schafer, F.; Badaan, S.; Vigaru, B.; Tseng, K.; Petrisor, D.; Trock, B.; Stoianovici, D. (2011). "Tandem-robot Assisted Laparoscopic Radical Prostatectomy to Improve the Neurovascular Bundle Visualization: A Feasibility Study." (PDF). *Urology* **77** (2): 502–6. doi:10.1016/j.urology.2010.06.064.

3.12   External links

- Zuse Institute Berlin (ZIB)
- Amira
- Amira videos

Chapter 4

ANIMAL (image processing)

ANIMAL (first implementation: 1988 - revised: 2004) is an interactive environment for image processing that is oriented toward the rapid prototyping, testing, and modification of algorithms. To create ANIMAL (AN IMage ALgebra), XLISP of David Betz was extended with some new types: sockets, arrays, images, masks, and drawables.

The theoretical framework and the implementation of the working environment is described in the paper "ANIMAL: AN IMage ALgebra".[1]

In the theoretical framework of ANIMAL a digital image is a boundless matrix with its *history*. However, in the implementation it is bounded by a rectangular region in the discrete plane and the elements outside the region have a constant value. The size and position of the region in the plane (focus) is defined by the coordinates of the rectangle. In this way all the pixels, including those on the border, have the same number of neighbors (useful in local operators, such as digital filters). Furthermore, pixelwise commutative operations remain commutative on image level, independently on focus (size and position of the rectangular regions). The *history* is a list which tracks the operations and parameters applied to the matrix. This mechanism is useful to document algorithms and generate new functions.

ANIMAL has been ported to R, a freely available language and environment for statistical computing and graphics. The new implementation is free and is used in a recent book[2] to illustrate the use of template matching techniques in computer vision (see the preface of the book code companion).

4.1 References

[1] R. Brunelli and C. M. Modena, "ANIMAL: AN IMage ALgebra", High Frequency, 1989, LVIII:3:255-259

[2] R. Brunelli, *Template Matching Techniques in Computer Vision: Theory and Practice*, Wiley, ISBN 978-0-470-51706-2, 2009 (*TM book*)

Chapter 5

AutoCollage 2008

AutoCollage 2008 is a Microsoft photomontage desktop application. The software creates a collage of representative elements from a set of images. It is able to detect faces and recognize objects.[1]

The software was developed by Microsoft Research labs in Cambridge, England and launched on September 4, 2008.

An update, named Microsoft Research AutoCollage 2008 version 1.1, was released on February, 2009. The software update adds the ability to select images for the AutoCollage, a richer integration with Windows Live Photo Gallery, support for network folders and the ability to define custom output sizes.

A new version, named Microsoft Research AutoCollage Touch 2009, was released on September 2009, and included by some OEMs on machines with Windows 7.

5.1 References

[1] "Introducing Microsoft AutoCollage 2008". *Softpedia.com*. Archived from the original on 6 September 2008. Retrieved 2008-09-06.

5.2 External links

- Description of Microsoft Research AutoCollage 2008 version 1.1

- AutoCollage blog

Chapter 6

Avizo (software)

Avizo (pronounce: 'a-VEE-zo') is a general-purpose commercial software application for scientific and industrial data visualization and analysis.

Avizo is developed by **FEI Visualization Sciences Group** and was originally designed and developed by the Visualization and Data Analysis Group at Zuse Institute Berlin (ZIB) under the name Amira. Avizo was commercially released in November 2007.

6.1 Overview

Avizo is a software application which enables users to perform interactive visualization and computation on 3D data sets. The Avizo interface is modelled on the visual programming. Users manipulate data and module components, organized in an interactive graph representation (called Pool), or in a Tree view. Data and modules can be interactively connected together, and controlled with several parameters, creating a visual processing network whose output is displayed in a 3D viewer.

With this interface, complex data can be interactively explored and analyzed by applying a controlled sequence of computation and display processes resulting into a meaningful visual representation and associated derived data.

6.2 Application Areas

Avizo has been designed to support different types of applications and workflows - from 2D and 3D image data processing to simulations −. It is a versatile and customizable visualization tool used in many fields:

- Scientific Visualization[1][2]

- Materials Research[3][4][5][6][7][8][9][10][11][12]

- Tomography,[13][14][15][16] Microscopy,[17][18] etc.

- Non Destructive Testing,[19][20][21] Industrial Inspection,[22][23] and Visual Inspection

- Computer-aided Engineering[24][25] and simulation data post-processing

- Porous Media Analysis [26][27][28][29][30][31][32][33][34]

- Civil Engineering [35]

- Seismic Exploration, Reservoir Engineering, Microseismic Monitoring, Borehole Imaging[36]

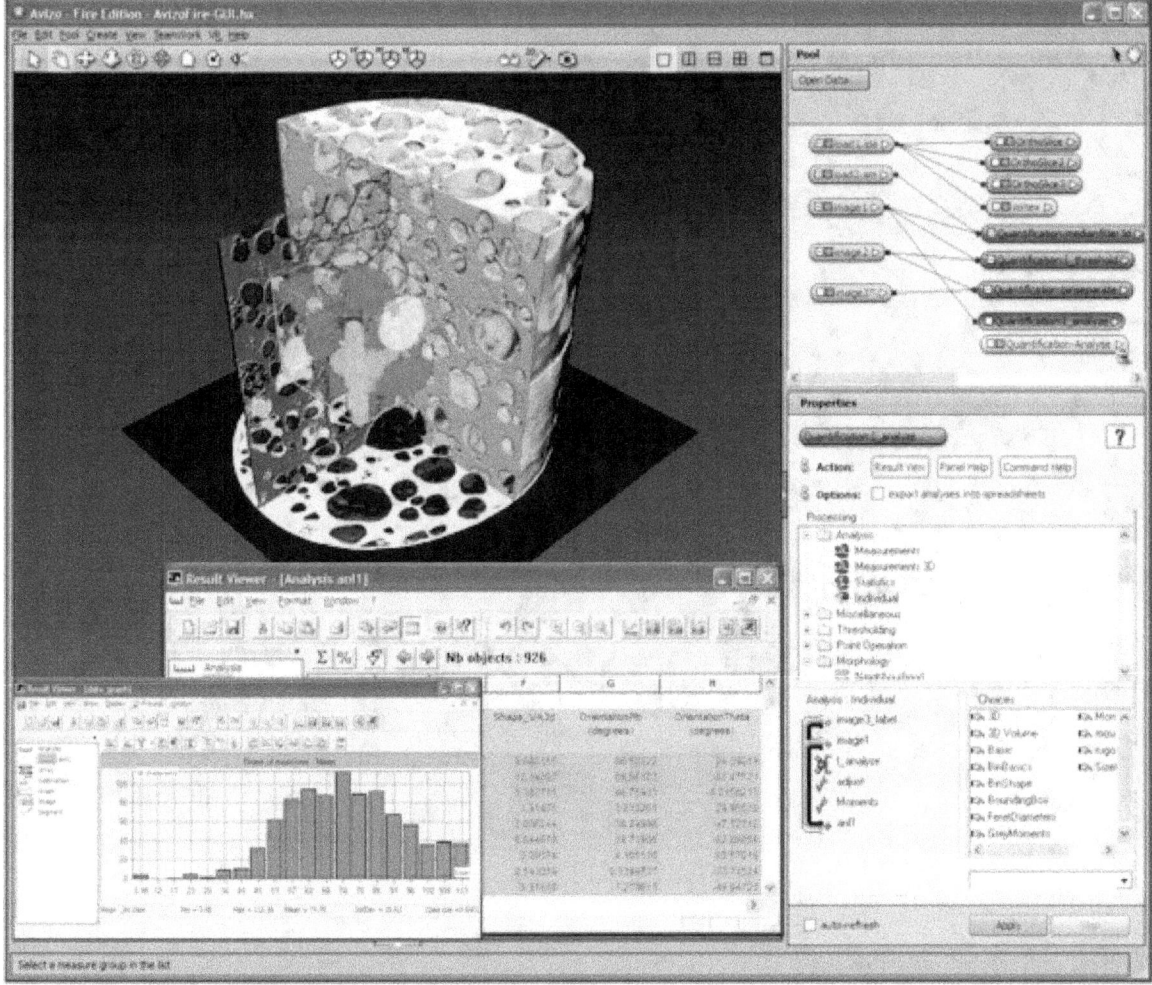

Metallic foam quantification

- Geology,[37][38][39] Digital Rock Physics (DRP),[40][41][42] Earth Sciences[43][44][45][46][47]

- Archaeology[48][49]

- Food Technology and Agricultural Science[50][51][52][53][54]

- Physics, Chemistry [55]

- Climatology,[56][57][58][59] Oceanography, Environmental Studies [60][61]

- Astrophysics

6.3   Features

Data Import:

- 2D and 3D image stack and volume data: from microscopes (electron, optical),[62][63][64][65][66][67][68] X-ray tomography (CT, micro-/nano-CT, synchrotron),[69][70][71][72] and other acquisition devices (MRI, radiography, GPR) [73]

- Geometric models (such as point sets, line sets, surfaces, grids)

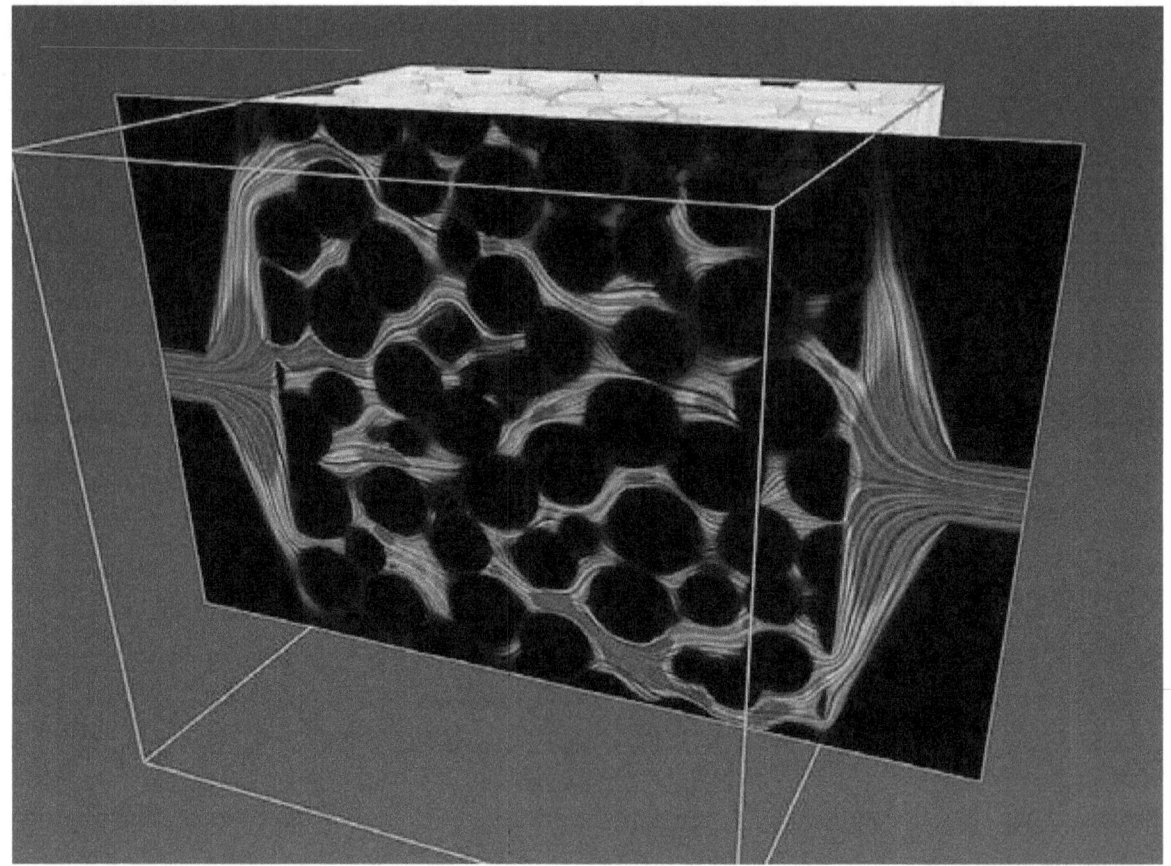

Virtual permeameter for absolute permeability computation

- Numerical simulation data [74][75][76][77] (such as Computational fluid dynamics or Finite element analysis data)

- Molecular data

- Time series and animations [78]

- Seismic data[79]

- Well logs

- 4D Multivariate Climate Models [80][81]

2D / 3D Data Visualization: [82][83][84]

- Volume rendering[85][86][87][88]

- Visualization of sections, through various slicing and clipping methods [89]

- Isosurface rendering[90][91][92][93][94]

- Polygonal meshes

- Scalar fields, Vector fields, Tensor representations, Flow visualization (Illuminated Streamlines, Stream Ribbons) [95]

Image Processing: [96][97][98][99][100][101][102][103]

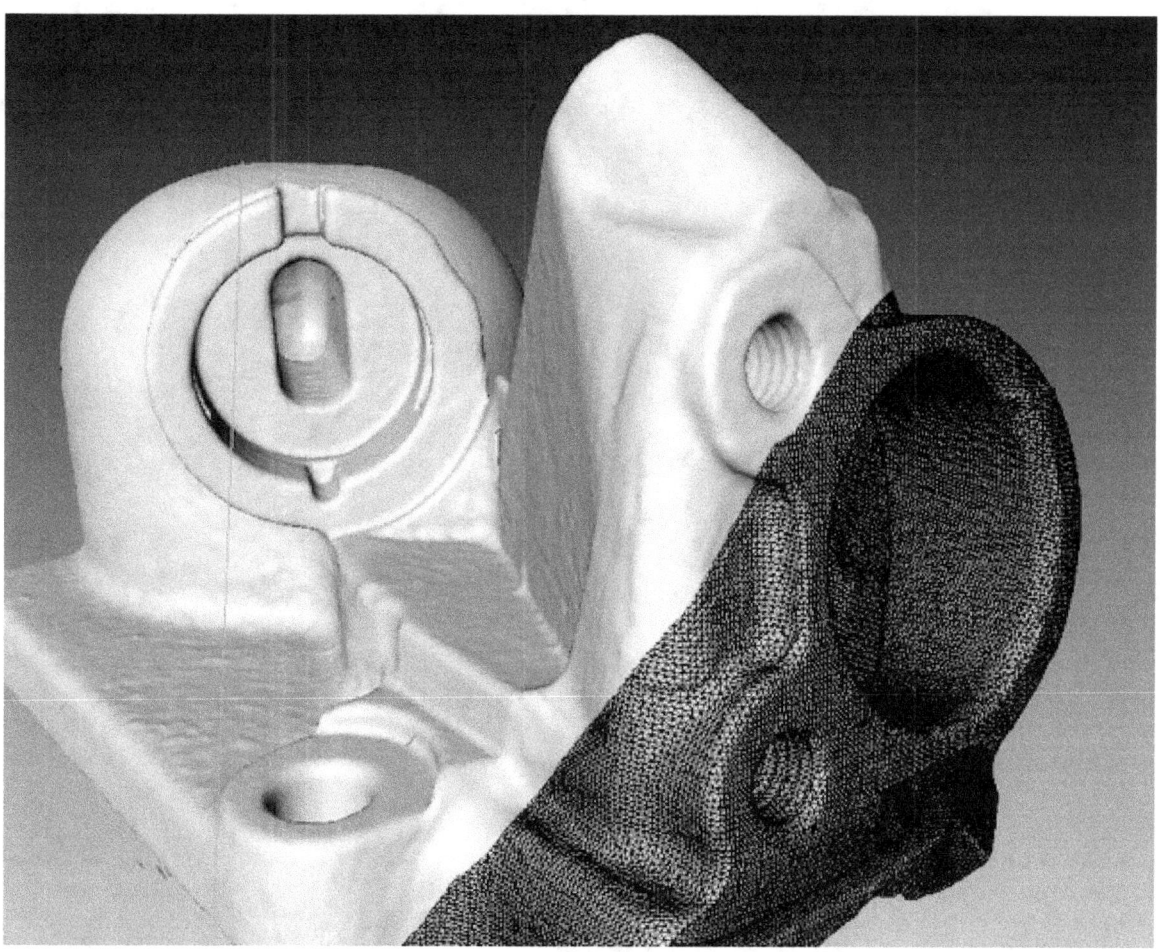

3D image-based meshing for CFD/FEA analysis of a mechanical part

- 2D/3D Alignment of image slices,[104] Image registration[105]

- Image filtering

- Mathematical Morphology (erode, dilate, open, close, tophat)

- Watershed Transform, Distance Transform

- Image segmentation[106][107][108][109][110][111][112][113][114][115][116]

3D Models Reconstruction: [117][118][119][120][121][122][123][124][125]

- Polygonal surface generation from segmented objects [126]

- Generation of tetrahedral grids [127]

- Surface reconstruction from point clouds

- Skeletonization (reconstruction of dendritic, porous or fracture network) [128][129][130][131][132]

- Surface model simplification

Quantification and analysis: [133][134][135][136][137][138][139][140][141][142]

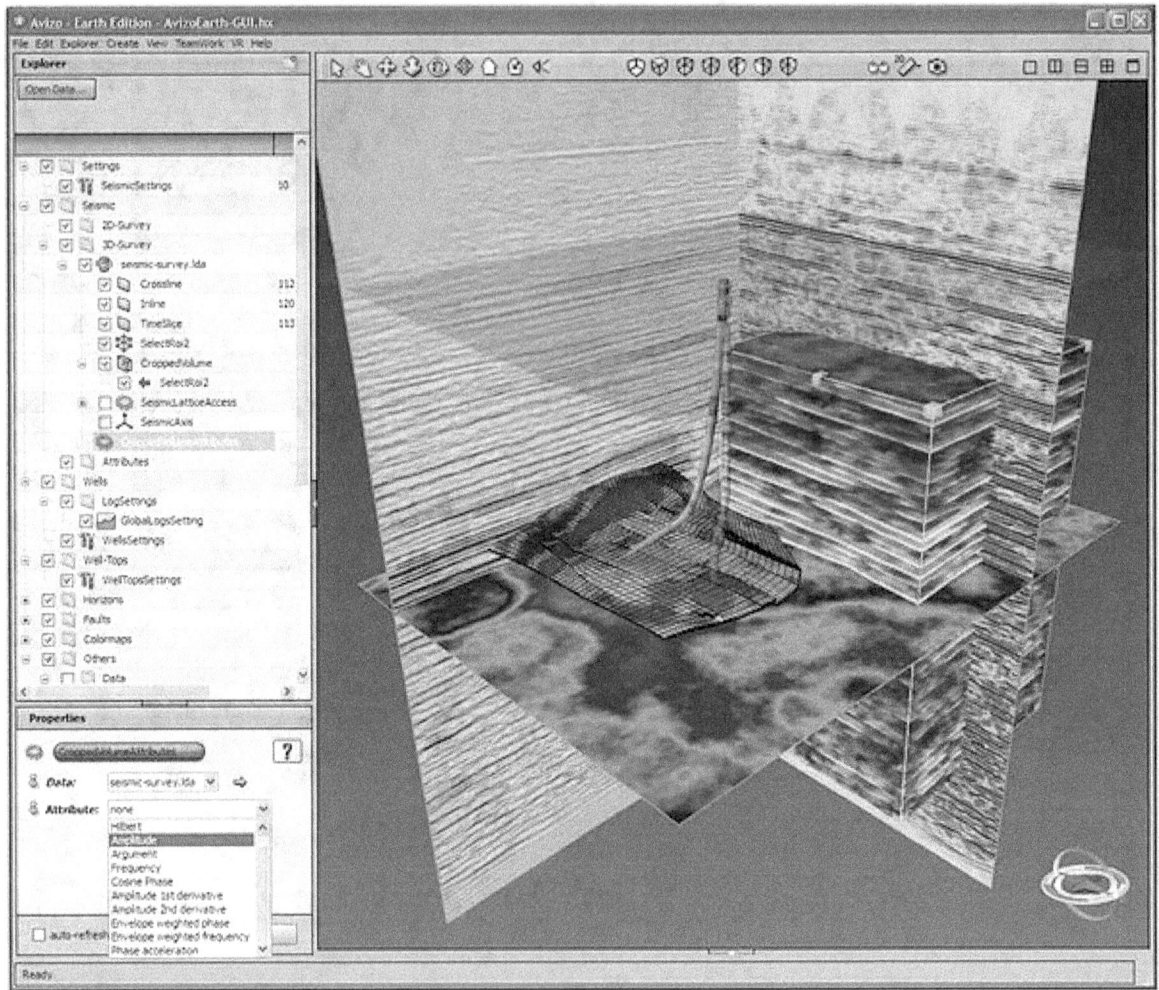

Geosciences data visualization

- Measurements and statistics [143][144][145][146][147][148][149]

- Analysis spreadsheet and charting

Material properties computation, based on 3D images:

- Absolute permeability

- Thermal conductivity

- Molecular diffusivity

- Electrical resistivity/Formation factor

3D image-based meshing for CFD and FEA: [150]

- From 3D imaging modalities (CT, micro-CT, MRI, etc.) [151][152]

- Surface and volume meshes generation [153]

- Export to FEA and CFD solvers for simulation

- Post-processing for simulation analysis

Presentation, Automation:

- MovieMaker,[154] Multiscreen, Video wall, collaboration, and VR[155] support
- TCL Scripting, C++ extension API

Avizo is based on Open Inventor 3D graphics toolkits (FEI Visualization Sciences Group).

6.4 External links

- Official website
- Scientific Publications
- Official Avizo forum
- Avizo videos

6.5 References

[1] LSU Honors Class 3035, Spring'11, Student Project Showcase - Louisiana State University, USA

[2] Maiolino S, Boyer DM, Bloch JI, Gilbert CC, Groenke J (2012) Evidence for a Grooming Claw in a North American Adapiform Primate: Implications for Anthropoid Origins. PLoS ONE 7(1): e29135. doi:10.1371/journal.pone.0029135

[3] Morris, R. A., Wang, B., Butts, D. and Thompson, G. B. (2012), Variations in Tantalum Carbide Microstructures with Changing Carbon Content. International Journal of Applied Ceramic Technology. doi:10.1111/j.1744-7402.2012.02761.x

[4] In situ observations of microscale damage evolution in unidirectional natural fibre composites, by Morten Rask (a), Bo Madsen (a), Bent F. Sørensen (a), Julie L. Fife (b), Karolina Martyniuk (a), Erik M. Lauridsen (a) - (a) Department of Wind Energy, Section of Composites and Materials Mechanics, Technical University of Denmark, Risø Campus (Denmark), (b) Swiss Light Source, Villigen (Switzerland)

[5] 3D Imaging and Metrology of Yttria Dispersoids in INCOLOY MA956 by Electron Tomography, by Adam Kruk et al., 2012, Solid State Phenomena, 186, 37

[6] Three-Dimensional Visualization and Metrology of Nanoparticles in Inconel 718 by Electron Tomography, by Krzysztof Kulawik et al., 2012, Solid State Phenomena, 186, 45

[7] 3D Imaging of Strengthening Particles in Cr-V-Mo (13HMF) Steel Using FIB/SEM Tomography, by Władysław Osuch et al., 2012, Solid State Phenomena, 186, 41

[8] On the interest of synchrotron X-ray imaging for the study of solidification in metallic alloys / De l'intérêt de l'imagerie X synchrotron pour l'étude de la solidification d'alliages métalliques, by Henri Nguyen-Thi (a, b), Luc Salvo (c), Ragnvald H. Mathiesen (d), Lars Arnberg (e), Bernard Billia (a, b), Michel Suery (c), Guillaume Reinhart (a, b) - (a) Aix Marseille University, IM2NP (France), (b) CNRS, IM2NP (France), (c) Université de Grenoble – CNRS, laboratoire SIMAP-GPM2 (France), (d) Department of Physics, NTNU, Trondheim (Norway), (e) Department of Materials Technology, NTNU, Trondheim (Norway)

[9] Relations entre la microstructure 3D et les propriétés mécaniques dans des réfractaires électrofondus à très haute teneur en zircone, by Yang Dinga (b), Michel Boussuge (a), Michel Gaubil (b), Samuel Forest (a), Ludovic Massard (b), Isabelle Cabodi (b), Sylvain Gailliègue (a) - (a) Centre des Matériaux, MINES-ParisTech, Evry (France), (b) Saint-Gobain CREE, Cavaillon (France)

[10] Topological Analysis of Martensite Morphology in Dual-Phase Steels, by N Sato, M Ojima, S Morooka, Y Tomota... - Advanced Materials Research, 2012

[11] Exploring capillary trapping efficiency as a function of interfacial tension, viscosity, and flow rate, (1) School of Chemical, Biological and Environmental Engineering, Oregon State University, Corvallis, Oregon, USA, (2) School of Environmental and Rural Sciences, University of New England, NSW, Australia, (3) Earth and Environmental Sciences Division (EES), Los Alamos National Laboratory, Los Alamos, New Mexico, USA.

[12] PRÉCIPITATION DU BORE DANS LE SILICIUM IMPLANTÉ ET REDISTRIBUTION DU BORE ET PLATINE LORS DE L'INTER-DIFFUSION RÉACTIVE DANS LES FILMS MINCES NICKEL/SILICIUM, by Oana Cojocaru-Mirédin (Rouen University, U.F.R. de Sciences et techniques, France)

[13] Microstructure and CO gas sensing property of Au/SnO2 core–shell structure nanoparticles synthesized by precipitation method and microwave–assisted hydrothermal synthesis method, by T Yanagimoto, YT Yu... - Sensors and Actuators B: Chemical, 2011

[14] Microbialite development patterns in the last deglacial reefs from Tahiti (French Polynesia; IODP Expedition #310): Implications on reef framework architecture, by Claire Seard (a), Gilbert Camoin (a), Yusuke Yokoyama (b, c), Hiroyuki Matsuzaki (d), Nicolas Durand (a), Edouard Bard (a), Sophie Sepulcre (a), Pierre Deschamps (a) - (a) CEREGE, Aix-Marseille Université, Aix-en-Provence (France), (b) Ocean Research Institute and Department of Earth and Planetary Sciences, University of Tokyo (Japan), (c) Institute for Research on Earth Evolution (IFREE), Japan Agency for Marine-Earth Science and Technology (JAMSTEC), Yokosuka (Japan), (d) Department of Nuclear Engineering and Management, University of Tokyo (Japan)

[15] APPLICATIONS OF THE CRACOW X-RAY MICROPROBE IN TOMOGRAPHY, by J. Bielecki, S. Bozek, J. Lekki, Z. Stachura and W.M. Kwiatek (The Henryk Niewodniczanski Institute of Nuclear Physics, Cracow, Poland)

[16] http://xinhuolin.web.officelive.com/default.aspx

[17] High temperature behavior of the metal/oxide interface of ferritic stainless steels, by Jérôme Issartel (a, b), Sébastien Martoia (b), Frédéric Charlot (c), Valérie Parry (a), Guillaume Parry (a), Rafael Estevez (a), Yves Wouters (a) - (a) SIMaP, Grenoble INP/CNRS/UJF (France), (b) Aperam Isbergues Research & Development (France), (c) CMTC, Grenoble INP/CNRS/UJF (France)

[18] 3-D illustration of network orientations of interstitial cells of Cajal subgroups in human colon as revealed by deep-tissue imaging with optical clearing, by Yuan-An Liu (1), Yuan-Chiang Chung (2), Shien-Tung Pan (3), Yung-Chi Hou(2), Shih-Jung Peng (1,4), Pankaj J. Pasricha (5), and Shiue-Cheng Tang(1,4) - (1) Department of Chemical Engineering, National Tsing Hua University; (2) Division of Colorectal Surgery, National Taiwan University Hospital–Hsinchu Branch; (3) Department of Pathology, National Taiwan University Hospital–Hsinchu Branch; (4) Institute of Biotechnology, National Tsing Hua University, Hsinchu, Taiwan; (5) Division of Gastroenterology and Hepatology, Stanford University School of Medicine, Stanford, California

[19] Non-destructive characterization of voids in six flowable composites using swept-source optical coherence tomography, by Amir Nazari(a, b, c), Alireza Sadr(d), Mohammad Ali Saghiri(c), Marc Campillo-Funollet(e), Hidenori Hamba(a), Yasushi Shimada(a), Junji Tagami(a), d, Yasunori Sumi(f) - (a) Graduate School of Medical and Dental Sciences, Tokyo Medical and Dental University (Japan), (b) NICOPE Corporation (Japan), (c) Kamal Asgar Research Center (KARC), Islamic Azad University (Iran), (d) International Research Center for Molecular Science in Tooth and Bone Diseases, Tokyo Medical and Dental University (Japan), (e) School of Dental Medicine, NY (USA), (f) National Hospital for Geriatric Medicine, National Center for Geriatrics and Gerontology (Japan)

[20] 3D Characterisation of Void Distribution in Resin Film Infused Composites - Fabien Léonard(1), Jasmin Stein(1,2), Arthur Wilkinson(2), Philip J. Withers(1), (1)Henry Moseley X-ray Imaging Facility, School of Materials, The University of Manchester, (2)Northwest Composites Centre, School of Materials, The University of Manchester

[21] Assessment by X-ray CT of the effects of geometry and build direction on defects in titanium ALM parts - Fabien Léonard(1), Samuel Tammas-Williams(1,2), Philip B. Prangnell(1), Iain Todd(2), Philip J. Withers(1), (1)Henry Moseley X-ray Imaging Facility, The University of Manchester, (2)University of Sheffield, Department of Materials Science and Engineering

[22] The role of cold work and applied stress on surface oxidation of 304 stainless steel, by S Lozano-Perez, K Kruska, I Iyengar, T Terachi... - Corrosion Science, 2011

[23] Effect of Porosity on the Fatigue Life of a Cast Al Alloy, By Nicolas Vanderesse (a), Jean-Yves Buffière (a), Eric Maire (a), Amaury Chabod (b) - (a) Université de Lyon - INSA de Lyon (France), (b) Centre Technique des Industries de la Fonderie (France)

[24] A VISUAL ANALYTICS APPROACH TO PRELIMINARY TRAJECTORY DESIGN, by Wayne R. Schlei, Kathleen C. Howell - Purdue University, School of Aeronautics and Astronautics (USA)

[25] DATA ACQUISITION FOR A BRIDGE COLLAPSE TEST, by Kurt Veggeberg (National Instruments) - Bridge Model Validation visualization with Avizo

[26] Upscaling Calcium Carbonate Precipitation Rates from Pore to Continuum Scale, by Catherine Noiriel (a) (b), Carl I. Steefel (b), Li Yang (b), Jonathan Ajo-Franklin (b) - (a) Géosciences Environnement Toulouse, UMR 5533 Université Paul Sabatier/CNRS/IRD/CNES, Toulouse (France), (b) Earth Sciences Division, Lawrence Berkeley National Laboratory, Berkeley, CA (USA) doi:10.1016/j.chemgeo.2012.05.014

[27] Wu, X., Schlangen, E. and van der Zwaag, S. (2012), Linking Porosity to Rolling Reduction and Fatigue Lifetime of Hot Rolled AA7xxx Alloys by 3D X-Ray Computed Tomography. Adv. Eng. Mater.. doi:10.1002/adem.201200087

[28] Evaluation of Asphalt Field Cores with Simple Performance Tester and X-ray Computed Tomography, by Florentina Angela Farca - Division of Highway and Railway Engineering, Department of Transport Science School of Architecture and the Built Environment, Royal Institute of Technology Stockholm (Sweden)

[29] Three-dimensional phase-field simulation of micropore formation during solidification: Morphological analysis and pinching effect, by H. Meidani (a), J.-L. Desbiolles (a), A. Jacot (a, b), M. Rappaz (a) - (a) Computational Materials Laboratory, Institute of Materials, Ecole Polytechnique Fédérale de Lausanne (Switzerland), (b) Calcom ESI SA, Lausanne (Switzerland)

[30] Three-Dimensional Simulation of SOFC Anode Polarization Characteristics Based on Sub-Grid Scale Modeling of Microstructure M Kishimoto, H Iwai, M Saito... - Journal of The Electrochemical Society, 2012

[31] 3D Quantitative Characterization of Nickel-Yttria-stabilized zirconia Solid Oxide Fuel Cell anode Microstructure in Discharge, by Z Jiao, N Shikazono...

[32] Poromechanics Investigation at Pore-scale Using Digital Rock Physics Laboratory, by S Zhang, N Saxena, P Barthelemy, M Marsh... - 2011

[33] X-ray tomography of catalyst layer of proton exchange membrane fuel cell, By YUHUNG HSU - The Henry Moseley X-ray Imaging Facility (HMXIF), School of Materials at the University of Manchester, UK

[34] Current status of imaging microbial biofilms in three-dimensional opaque porous media using x-ray microtomography, by Dorthe Wildenschild (1), Gabriel Iltis (1), Ryan Armstrong (1), Yohan Davit (2), James Connolly (3), Robin Gerlach(3), and Brian Wood (1) - (1) School of Chemical, Biological and Environmental Engineering, Oregon State University, (2) Institut de Mecanique des Fluides de Toulouse, Universite de Toulouse, France, (3) Center for Biofilm Engineering Montana State University.

[35] Three-dimensional microstructural modeling of asphalt concrete using a unified viscoelastic–viscoplastic–viscodamage model, by T You, RK Abu Al-Rub, MK Darabi, EA Masad... - Construction and Building ..., 2012

[36] CHARACTERIZATION OF HETEROGENEITIES FROM CORE X-RAY SCANS AND BOREHOLE WALL IMAGES IN A REEFAL CARBONATE RESERVOIR: INFLUENCE ON THE POROSITY STRUCTURE, by V. Hebert, C. Garing, P.A. Pezard, P. Gouze, and Y. Maria-Sube (Geosciences Montpellier, CNRS, University of Montpellier 2, Montpellier, France), G. Camoin (CEREGE, CNRS, Aix-en-Provence, France), and P. Lapointe (TOTAL - CSTJF, Pau, France)

[37] 4D imaging of fracturing in organic-rich shales during heating, by PGP in JOURNAL OF GEOPHYSICAL RESEARCH, VOL. 116, B12201, 9 PP., 2011 doi:10.1029/2011JB008565

[38] Fossilized fungi in subseafloor Eocene basalts - Magnus Ivarsson (1), Stefan Bengtson (1), Veneta Belivanova (2), Marco Stampanoni (3,4), Federica Marone (3) and Anders Tehler (5) - (1) Department of Palaeozoology and Nordic Center for Earth Evolution, Swedish Museum of Natural History, Stockholm, Sweden; (2) Department of Palaeozoology, Swedish Museum of Natural History, Stockholm, Sweden; (3) Swiss Light Source, Paul Scherrer Institute, Villigen, Switzerland; (4) Institute for Biomedical Engineering, University and ETH Zürich, Switzerland; (5) Department of Cryptogamic Botany, Swedish Museum of Natural History, Stockholm, Sweden

[39] Pore formation during dehydration of polycrystalline gypsum observed and quantified in a time-series synchrotron radiation based X-ray micro-tomography experiment, by F. Fusseis (1,2,3), C. Schrank (1,2,3), J. Liu (1,2), A. Karrech (1,4), S. Llana-Funez (5), X. Xiao (6), and K. Regenauer-Lieb (1,2,3,4) - (1) Multi-scale Earth System Dynamics, Perth, Australia, (2) School of Earth and Environment, University of Western Australia, Crawley, Australia, (3) Western Australian Geothermal Centre of Excellence, Perth, Australia, (4) CSIRO Earth Science and Resource Engineering, Kensington, Australia, (5) Departamento de Geologia, Universidad de Oviedo, Oviedo, Spain

[40] Shulakova, V., Pervukhina, M., Müller, T. M., Lebedev, M., Mayo, S., Schmid, S., Golodoniuc, P., De Paula, O. B., Clennell, M. B. and Gurevich, B. (2012), Computational elastic up-scaling of sandstone on the basis of X-ray micro-tomographic images. Geophysical Prospecting. doi:10.1111/j.1365-2478.2012.01082.x

[41] QUANTITATIVE CHARACTERIZATION OF 3D MELT DISTRIBUTION IN PARTIALLY MOLTEN OLIVINE6BASALT AGGREGATES USING X-RAY SYNCHROTRON MICROTOMOGRAPHY, by Zhu W., Gaetani G. A., Fusseis F. - American Geophysical Union, Fall Meeting 2009, abstract #V33A-2029

[42] IMPROVED METHODOLOGY FOR THE CHARACTERIZATION OF COMPLEX VUGGY CARBONATE, by A. Bersani, B.B. Bam, F. Radaelli and E. Rossi (ENI E&P)

[43] 3-D imaging and quantification of graupel porosity by synchrotron-based micro-tomography, by F. Enzmann (1), M. M. Miedaner (1), M. Kersten (1), N. von Blohn (1), K. Diehl (1), S. Borrmann (1), M. Stampanoni (2), M. Ammann (2), and T. Huthwelker (2) - (1) Earth System Science Research Centre, Johannes Gutenberg-University, Mainz (Germany), (2) Paul Scherrer Institut, Villigen-PSI, Villigen (Switzerland)

[44] X-Ray Microtomography for Studying 3D Textures of Speleothems Developed inside Historic Walls, by JAVIER MARTÍNEZ-MARTÍNEZ (1,2,*), NICOLETTA FUSI (3), VALENTINA BARBERINI (3), JUAN CARLOS CAÑAVERAS(1,2), GIOVANNI BATTISTA CROSTA (3) - (1) Laboratorio de Petrología Aplicada. Universidad de Alicante (Spain), (2) Dep. Ciencias de la Tierra y del Medio Ambiente. Universidad de Alicante (Spain), (3) Dip. Scienze Geologiche e Geotecnologie. Università degli Studi di Milano–Bicocca (Italia)

[45] PHYSICAL AND MECHANICAL CHARACTERIZATION OF ALTERED VOLCANIC ROCKS FOR THE STABILITY OF VOLCANIC EDIFICES, by Dr. Antonio Pola Villaseñor - Università degli Studi di Milano-Bicocca (Facoltà di Scienze Matematiche, Fisiche, Naturali Dipartimento di Scienze Geologiche e Geotecnologie)

[46] DISCRETE 3D DIGITAL IMAGE CORRELATION (DIC) USING PARTICLE RECONNAISSANCE, by Rutger C.A. Smit, January 7, 2010

[47] THE CONTROVERSIAL "CAMBRIAN" FOSSILS OF THE VINDHYAN ARE REAL BUT MORE THAN A BILLION YEARS OLD, by Stefan Bengtson(a)(b)(1), Veneta Belivanova(a), Birger Rasmussen(c), and Martin Whitehouse(b)(d) - (a) Department of Palaeozoology and (d) Laboratory for Isotope Geology, Swedish Museum of Natural History, SE-104 05 Stockholm, Sweden;(b) Nordic Center for Earth Evolution, SE-104 05 Stockholm, Sweden; and (c) Department of Applied Geology, Curtin University of Technology, Perth, WA 6845, Australia

[48] VISTA Center, Institute of Archaeology and Antiquity, University of Birmingham, UK

[49] MAPPING DOGGERLAND: The Mesolithic Landscapes of the Southern North Sea. Archaeopress. Oxford. Gaffney V., Thomson K. and Fitch S. (Eds.) 2007

[50] Modeling of mass transfer and initiation of hygroscopically induced cracks in rice grains in a thermally controlled soaking condition: With dependency of diffusion coefficient to moisture content and temperature – A 3D finite element approach, by Jonathan H. Perez (a), Fumihiko Tanaka (b), Toshitaka Uchino (b) - (a) Graduate School of Bioresource and Bioenvironmental Sciences, Kyushu University (Japan), (b) Faculty of Agriculture, Kyushu University (Japan)

[51] DIFFERENCES IN THREE-DIMENSIONAL STRUCTURE REVEALED BY HIGH RESOLUTION MICRO X-RAY TOMOGRAPHY ARE RELATED TO FRESH AND COOKED MEAT TENDERNESS, by VLAD BRUMFELD* AND DAVID E. GERRARD - Electron Microscopy Unit, The Weizmann Institute of Science Rehovot, ISRAEL AND Department of Animal & Poultry Sciences Virginia Tech USA

[52] Viscoelastic stress-development during drying of corn kernels: a multiscale porous media study from cellular to tissue scales, P S. Takhar, Department of Animal and Food Sciences, International Center for Food Industry Excellence, Texas Tech University, Lubbock, TX (USA)

[53] High-resolution micro- and nano-CT of soft food materials: InsideFood, by E. Herremans (1), B. E. Verlinden (2), E.Bongaers (3), B. Pauwels (3), E. Jakubczyk (4), P. Estrade(5), P. Verboven(1), B.M.Nicolaï (1,2) - (1) BIOSYST-MeBioS, K.U.Leuven (Belgium), (2) VCBT, Flanders Centre of Postharvest Technology (Belgium), (3) SkyScan (Belgium), (4) SGGW, Warsaw University of Life Sciences, Dep. Food Eng. & Process Management (Poland), (5) VSG, Visualization Sciences Group SAS (France)

[54] Journey to the centre of an apple, by MeBios Biofluidics Group and the Flemish Primitives

[55] Exploring microstructural changes associated with oxidation in Ni–YSZ SOFC electrodes using high resolution X-ray computed tomography, by PR Shearing, RS Bradley, J Gelb, F Tariq… - Solid State Ionics, 2011

[56] Neset, T-S S., Johansson, J. and Linnér, B-O. (eds.) (2009). STATE OF CLIMATE VISUALISATION, CSPR Report N:o 09:04, Centre for Climate Science and Policy Research, Norrköping, Sweden

[57] 2008 Geoinformatics Conference - AVIZO - 3D VISUALIZATION FRAMEWORK -, by Peter Westenberger

[58] INTERACTIVE DATA VISUALIZATION WITH FOCUS ON EARTH SYSTEM RESEARCH, by Michael Böttinger (DKRZ, German Climate Computing Centre)

[59] Simulation of typhoon Morakot (2009), Chinese winter storms (2008) and hurricane Katrina (2005), By Taiwan Typhoon and Flood Research Institute (TTFRI)

[60] Infrared thermography of evaporative fluxes and dynamics of salt deposition on heterogeneous porous surfaces, by U Nachshon, E Shahraeeni, D Or, M Dragila... - Water Resources Research, 2011

[61] Multi-scale Simulation of tsunami from Tohoku Earthquake and Diffracted Tsunami Waves in Hokkaido, By ZHU AIYU (1), Yuen Dave A. (2), Song Shenyi (3) - (1) Zhang Dongning Institute of Geophysics, China Earthquake Administration, (2) University of Minnesota, (3) Computer Network Information Center, China Academy of Science [www.es.ucsc.edu/~{}acti/sanya/Aiyu.ppt]

[62] TOPOLOGY OF SPHEROIDIZED PEARLITE, by Yoshitaka Adachi (1) and Yuan Tsung Wang (1) - (1) National Institute for Materials Science, Tsukuba, Japan

[63] Evaluation of fracture toughness of small volumes by means of cube-corner nanoindentation, by N. Cuadrado (a), D. Casellas (a, b), M. Anglada (c), E. Jiménez-Piqué (c) - (a) Fundació CTM Centre Tecnològic, Manresa (Spain), (b) Department of Materials Science and Metallurgical Engineering, EPSEM, Universitat Politècnica de Catalunya, Manresa (Spain), (c) Department of Materials Science and Metallurgical Engineering, ETSEIB, Universitat Politècnica de Catalunya, Barcelona (Spain)

[64] POROSITY AND PERMEABILITY ANALYSIS ON NANOSCALE FIB-SEM 3D IMAGING OF SHALE ROCK, By Shawn Zhang (1), Robert E. Klimentidis (2), Patrick Barthelemy (1) - (1) Visualization Sciences Group (VSG), (2) Exxon-Mobil Upstream Research Co.

[65] FIB Tomographic Analysis of Subsurface Indentation Crack Interactions with Pores in Alumina, by J. A. Arsecularatne, M. Hoffman, K. O'Kelly, N. Payraudeau - In Journal of the American Ceramic Society, Volume 94, Issue 11, pages 4017–4024, November 2011

[66] N. Vivet, S. Chupin, E. Estrade, T. Piquero, P.L. Pommier, D. Rochais, E. Bruneton, 3D Microstructural characterization of an SOFC anode reconstructed by FIB tomography, Journal of Power Sources (2010), doi:10.1016/j.jpowsour.2011.03.060

[67] FIB SEQUENTIAL SECTIONING AS A TOOL TO UNDERSTAND ENVIRONMENTAL MATERIALS DEGRADATION IN 3D 環境材料劣化を3 次元で理解するためのツールとしての FIB による連続断面試料作製観察方法, by S. Lozano-Perez, NNi, K. Kruska, T. Terachi, and T. Yamada

[68] PHASE CONTINUITY IN HIGH TEMPERATURE Mo-Si-B ALLOYS: A FIB-TOMOGRAPHY STUDY, by O. Hassomeris (a), G. Schumacher (a)*, M. Krüger (b), M. Heilmaier (c), J. Banhart (a) - (a) Helmholtz-Zentrum Berlin für Materialien und Energie GmbH, Berlin, Germany; (b) Otto-von-Guericke Universität Magdeburg, Institute for Materials and Joining Technology, Germany; (c) TU Darmstadt, Materials Science FB 11, Darmstadt, Germany

[69] CrossBeam Nano-Tomography & 3D Analytics in Research, By Dr. Daniel Kraft - Carl Zeiss SMT

[70] Virtual taphonomy using synchrotron tomographic microscopy reveals cryptic features and internal structure of modern and fossil plants, by Selena Y. Smith (a), Margaret E. Collinson (a), Paula J. Rudall (b), David A. Simpson (c), Federica Marone (d) and Marco Stampanonid (e) - (a) Department of Earth Sciences, Royal Holloway University of London (UK), (b) Jodrell Laboratory, Royal Botanic Gardens (UK), (c) Herbarium, Royal Botanic Gardens (UK), (d) Swiss Light Source, Paul Scherrer Institut (Switzerland), (e) Institute for Biomedical Engineering, University and Eidgenössische Technische Hochschule Zurich (Switzerland)

[71] DETERMINATION OF BUBBLES IN FOODS BY X-RAY MICROTOMOGRAPHY AND IMAGE ANALYSIS, by G. van Dalen, A. Don, P. Nootenboom, J.C.G. Blonk - Unilever R&D (The Netherlands)

[72] 3D SYNCHROTRON X-RAY MICROTOMOGRAPHY OF PAINT SAMPLES, Ester S.B. Ferreira (1), Jaap J. Boon (1) (2), Jerre van der Horst (2), Nadim C. Scherrer (1)(3), Federica Marone (4) and Marco Stampanoni (4) - (1) SIK-ISEA, Switzerland; (2) FOM Institute AMOLF, Netherlands; (3) Bern University of Applied Sciences, Switzerland; (4) TOMCAT, Swiss Light Source, Paul Scherrer Institute, Switzerland -

[73] SEM TECHNOLOGY ADVANCES ENERGY RESEARCH, by Natasha Erdman, Naoki Kikuchi, Regina Campbell, Vernon E. Robertson (JEOL USA Inc.)

[74] VISUALIZATION AND ANALYSIS OF THE GREAT SPHINX EROSION, by the VISTA team of Bibliotheca Alexandrina (Cairo, Egypt)

[75] COMSOLConference 2009: Drying of Corn Kernels: From Experimental Images to Multiscale Multiphysics Modeling, by Pawan S. Takhar (Texas Tech University) and Shuang Zhang (Visualization Sciences Group), 2009

[76] ETUDE DU REFROIDISSEMENT POSTCOULEE DE REFRACTAIRES ELECTROFONDUS RICHES EN ZIRCONE : OPTIMISATION DES PROCEDES ET DES MICROSTRUCTURES, by L. Laurence, Y. Ding, M. Boussuge, D. Ryckelynck, S. Forest, 8th Sept 2009

[77] NATIONAL INSTRUMENTS - ENHANCED 3D VISUALIZATION USING THE LABVIEW INTERFACE FOR VSG AVIZO -

[78] ONE-YEAR ANIMATED PLANET SIMULATOR CIRCULATION, by Meteorologisches Institut Universität Hamburg (Germany)

[79] Visual and Spatial Technology Center (VISTA) (University of Birmingham, UK)

[80] INTERACTIVE 3D VISUALIZATION IN EARTH SYSTEM RESEARCH, by The German Climate Computing Center (Deutsches Klimarechenzentrum, DKRZ)

[81] 3D-VISUALISIERING VON DATEN AUS KLIMAMODELLEN, by Carmen Ulmen, Dipl.-Geografin, Akad. Geoinformatikerin KlimaCampus Hamburg, Climate System Analysis and Prediction (CliSAP)

[82] Fossil jawless fish from China foreshadows early jawed vertebrate anatomy, by Zhikun Gai (1,2), Philip C. J. Donoghue (1), Min Zhu (2), Philippe Janvier (3), Marco Stampanoni (4,5) - (1) School of Earth Sciences, University of Bristol, Bristol (UK), (2) Laboratory of Evolutionary Systematics of Vertebrates, Institute of Vertebrate Paleontology and Paleoanthropology, Chinese Academy of Sciences, Beijing (China), (3) Muséum National d'Histoire Naturelle, Paris (France), (4) Swiss Light Source, Paul Scherrer Institut, Villigen (Switzerland), (5) Institute for Biomedical Engineering, University and ETH Zurich (Switzerland)

[83] Volume Edit and Thresholding - By Elise Orellana (LSU Honors Class 3035, Spring'11, Student Project Showcase) The Use of Metal Markers In 3D Imaging - By Caroline Blevins (LSU Honors Class 3035, Spring'11, Student Project Showcase) Using 3D Reconstruction to Illustrate the Cranio-cervical Envelope in the Alligator - By Brooke Hopkins (LSU Honors Class 3035, Spring'11, Student Project Showcase) Visualization and Analysis of the Human Shoulder Suspensory Apparatus - By Michelle Osborn (LSU Honors Class 3035, Spring'11, Student Project Showcase)

[84] 3D Quantification of Plant Root Architecture In Situ, By Suqin Fang, Randy Clark and Hong Liao - Measuring Roots, 2011

[85] Herrel, A., Fabre, A.-C., Hugot, J.-P., Keovichit, K., Adriaens, D., Brabant, L., Van Hoorebeke, L. and Cornette, R. (2012), Ontogeny of the cranial system in Laonastes aenigmamus. Journal of Anatomy. doi:10.1111/j.1469-7580.2012.01519.x

[86] Helical arrangement of filaments in microvillar actin bundles, by Keisuke Ohta (a), Ryuhei Higashi (b), Akira Sawaguchi (c), Kei-ichiro Nakamura (a) - (a) Division of Microscopic and Developmental Anatomy, Department of Anatomy, Kurume University School of Medicine, Fukuoka (Japan), (b) Electron Microscopic Laboratory, Central Research Unit of Kurume University, Fukuoka (Japan), (c) Department of Anatomy, Ultrastructural Cell Biology, Faculty of Medicine, University of Miyazaki, Miyazaki (Japan)

[87] VISUALISING THE 3D INTERNAL STRUCTURE OF CALCITE SINGLE CRYSTALS GROWN IN AGAROSE HYDROGELS, by Hanying Li (1) Huolin L. Xin (2) David A. Muller (3) Lara A. Estroff (1) - (1) Department of Materials Science and Engineering, (2) Department of Physics, (3) School of Applied and Engineering Physics, Cornell University (NY, USA)

[88] LARGE COLONIAL ORGANISMS WITH COORDINATED GROWTH IN OXYGENATED ENVIRONMENTS 2.1 Gye ago, by Abderrazak El Albani (1), Stefan Bengtson (2), Donald E. Canfield (3), Andrey Bekker (4), Roberto Macchiarelli (5)(6), Arnaud Mazurier (7), Emma U. Hammarlund (2)(3)(8), Philippe Boulvais (9), Jean-Jacques Dupuy (10), Claude Fontaine (1), Franz T. Fursich (11), François Gauthier-Lafaye (12), Philippe Janvier (13), Emmanuelle Javaux (14), Frantz Ossa Ossa (1), Anne-Catherine Pierson-Wickmann (9), Armelle Riboulleau (15), Paul Sardini (1), Daniel Vachard (15), Martin Whitehouse (16) & Alain Meunier (1) - (1)Laboratoire HYDRASA, Université de Poitiers, France. (2)Department of Palaeozoology, Swedish Museum of Natural History, Stockholm. (3)Nordic Center for Earth Evolution, Denmark. (4)Department of Geological Sciences, University of Manitoba, Canada. (5)Département Géosciences, Centre de Microtomographie, Université de Poitiers, France. (6)Département de Préhistoire, Muséum National d'Histoire Naturelle, Paris, France. (7)Société Etudes Recherches Matériaux, Poitiers, France. (8)Department of Geological Sciences, Stockholm University, Sweden. (9)Département Géosciences, Université de Rennes, France. (10)Bureau de Recherches Géologiques et Minières, Orléans, France. (11)GeoZentrum Nordbayern, Universitat Erlangen, Germany. (12)Laboratoire d'Hydrologie et de Géochimie de Strasbourg,

France. (13)Département Histoire de la Terre, Muséum National d'Histoire Naturelle, Paris, France. (14)Département de Géologie, Université de Liège, Belgium. (15)Laboratoire Géosystèmes, Université de Lille, France. (16)Laboratory for Isotope Geology, Swedish Museum of Natural History, Stockholm.

[89] Characterization Of Nanodevices By STEM Tomography, by O Richard, A Vandooren, GS Kar, P Van Marcke… - AIP Conf. Proc, 2011

[90] Differential attraction and repulsion of Staphylococcus aureus and Pseudomonas aeruginosa on molecularly smooth titanium films, by EP Ivanova, VK Truong, HK Webb, VA Baulin… - Scientific Reports, 2011 WT.mc_id=FBK_SciReports

[91] Mineral and Matrix Components of the Operculum and Shell of the Barnacle Balanus amphitrite: Calcite Crystal Growth in a Hydrogel, by Gal Mor Khalifa, Steve Weiner, and Lia Addadi - Department of Structural Biology, Weizmann Institute of Science, Rehovot (Israel)

[92] 3D IMAGING OF CEREAL FOOD PRODUCTS USING X-RAY MICROTOMOGRAPHY AND SEGMENTATION OF PHASE CONTRAST IMAGES, by G. van Dalen (1), A. Don (1), P. Nootenboom (1), M. van Ginkel (2), E. Boller (3), M. Langer (3) - (1)Unilever R&D, Vlaardingen, the Netherlands; (2)Unilever R&D, Colworth, UK; (3)European Synchrotron Radiation Facility, Grenoble, France

[93] CREEP CAVITATION CAN ESTABLISH A DYNAMIC GRANULAR FLUID PUMP IN DUCTILE SHEAR ZONES, by F. Fusseis (1), K. Regenauer-Lieb(1)(2), J. Liu(2), R. M. Hough(2) & F. De Carlo(3) - (1) School of Earth & Environment, The University of Western Australia. (2) CSIRO Exploration & Mining. (3) Advanced Photon Source, Argonne National Laboratory

[94] FGF/FGFR Signaling Coordinates Skull Development by Modulating Magnitude of Morphological Integration: Evidence from Apert Syndrome Mouse Models, By Neus Martínez-Abadías (1), Yann Heuzé (1), Yingli Wang (2), Ethylin Wang Jabs (2), Kristina Aldridge (3), Joan T. Richtsmeier (1) - (1) Department of Anthropology, Pennsylvania State University (USA), (2) Department of Genetics and Genomic Sciences, Mount Sinai School of Medicine, New York (USA), (3) Department of Pathology and Anatomical Sciences, University of Missouri-School of Medicine, Columbia, Missouri (USA)

[95] SIMULATION OF CELL SEEDING AND RETENTION IN A DISORDERED POLYMERIC SCAFFOLD, by Tejaswini Narayana, University of Kerala (India)

[96] Structural Characterization of RDX-based Explosive Nanocomposites - Stepanov, V., Willey, T. M., Ilavsky, J., Gelb, J. and Qiu, H. (2012), Structural Characterization of RDX-based Explosive Nanocomposites. Propellants, Explosives, Pyrotechnics. doi:10.1002/prep.201200151

[97] Tooth Movements are Guided by Specific Contact Areas Between the Tooth Root and the Jaw Bone: a Dynamic 3D Micro-CT Study of the Rat Molar, by GRS Naveh, R Shahar, V Brumfeld… - Journal of Structural Biology, 2011

[98] Direct Pore-scale Modeling of Two-phase Flow Through Natural Media, by I Bogdanov, F Guerton, J Kpahou…

[99] Fronto-facial monobloc distraction in syndromic craniosynostosis. Three-dimensional evaluation of treatment outcome and facial growth, by EWC Ko, PKT Chen, ICH Tai… - … Journal of Oral and Maxillofacial Surgery, 2011

[100] Beam deceleration for block-face scanning electron microscopy of embedded biological tissue, by Keisuke Ohta (a), Shoji Sadayama (b,c), Akinobu Togo (d), Ryuhei Higashi (d), Ryuichiro Tanoue (a), Kei-ichiro Nakamura (a) - (a) Division of Microscopic and Developmental Anatomy, Kurume University school of Medicine, Fukuoka (Japan), (b) Department of Materials science and Engineering, Kyusyu University, Fukuoka (Japan), (c) Application Laboratory, FEI Company Japan Ltd., Tokyo (Japan), (d) Electron Microscopic Laboratory, Central Research Unit of Kurume University, Fukuoka (Japan)

[101] IN SITU high speed X-ray observation of the solidification of AL15CU with and without AL2O3 composite addition, By R.W. Hamilton (1), A. Leung (1), A.B. Phillion (2), P. Rockett (1), T. Connolley (3), and P.D. Lee (1) - (1) Department of Materials, Imperial College London (UK), (2) School of Engineering, The University of British Columbia (Canada), (3) Diamond Light Source Ltd (UK) - in Shape Casting: Fourth International Symposium 2011 (in Honor of Prof. John T. Berry)

[102] Whispering to the Deaf: Communication by a Frog without External Vocal Sac or Tympanum in Noisy Environments, by Renaud Boistel (1)(2), Thierry Aubin (1), Peter Cloetens (3), Max Langer (3)(4), Brigitte Gillet (5), Patrice Josset (6), Nicolas Pollet(7), Anthony Herrel (8) - (1) Centre de Neurosciences Paris-Sud (CNPS), Centre National de la Recherche Scientifique (CNRS), Université Paris XI, France; (2) Institut International de Paléoprimatologie et de Paléontologie Humaine (IPHEP), Centre National de la Recherche Scientifique (CNRS), Université de Poitiers, France; (3) European Synchrotron Radiation Facility, Grenoble, France; (4) Centre de Recherche en Acquisition et Traitement de l'Image pour la Santé (CREATIS), Centre National de la Recherche Scientifique (CNRS), Institut National de la Santé et de la Recherche Médicale (INSERM), Université

Claude Bernard Lyon 1, Institut National des Sciences Appliquées Lyon, France, (5) Imagerie par Résonance Magnétique Médicale et Multi-Modalités (IR4M), Centre National de la Recherche Scientifique (CNRS), Université Paris-Sud, France; (6) Hôpital d'Enfants Armand Trousseau, Paris, France; (7) Institute of Systems and Synthetic Biology, Genopole, Centre National de la Recherche Scientifique (CNRS), University of Evry, France; (8) Département Ecologie et Gestion de la Biodiversité (EGB), Centre National de la Recherche Scientifique (CNRS), Muséum National d'Histoire Naturelle, Paris, France.

[103] THREE-DIMENSIONAL PORE SCALE FLUID FLOW SIMULATION BASED ON COMPUTED MICROTOMOGRA-PHY CARBONATE ROCKS' IMAGES, by Jan Kaczmarczyk, Marek Dohnalik, Jadwiga Zalewska - Oil and Gas Institute, Well Logging Dpt, Kraków, Poland

[104] Teeth before jaws? Comparative analysis of the structure and development of the external and internal scales in the extinct jawless vertebrate Loganellia scotica, by M Rücklin, S Giles, P Janvier... - Evolution & Development, 2011

[105] On the use of peak-force tapping atomic force microscopy for quantification of the local elastic modulus in hardened cement paste - By Pavel Trtik (a), Josef Kaufmann (a), Udo Volz (b), (a) Empa, Swiss Federal Laboratories for Materials Science and Technology (Switzerland), (b) Bruker Nano GmbH (Germany)

[106] High-resolution three-dimensional reconstruction of a whole yeast cell using focused-ion beam scanning electron microscopy - In BioTechniques, Vol. 53, No. 1, July 2012 - By Dongguang Wei (1), Scott Jacobs (2), Shannon Modla (2), Shuang Zhang (3), Carissa L. Young (4), Robert Cirino (2), Jeffrey Caplan (2), and Kirk Czymmek (2, 5) - (1) Carl Zeiss Microscopy, NY, USA, (2) UD Bio-Imaging Center, Delaware Biotechnology Institute, University of Delaware, DE, USA, (3) Visualization Sciences Group, MA, USA, (4) Department of Chemical and Biomolecular Engineering, University of Delaware, DE, USA, (5) Department of Biological Sciences, University of Delaware, DE, USA

[107] 3-Dimensional Modeling of Graphitic foam Heat Sink, by A Bradu... - ... in Strategic Materials and Computational Design II

[108] Matching 4D Porous Media Fluid Flow GeoPET Data With COMSOL Multiphysics Simulation Results, by J Lippmann-Pipke, J Kulenkampff, G Marion...

[109] Locating the Acupoint Baihui (GV20) Beneath the Cerebral Cortex with MRI Reconstructed 3D Neuroimages, by Ein-Yiao Shen (1, 2), Fun-Jou Chen (3), Yun-Yin Chen (1) and Ming-Fan Lin (4) - (1) Graduate Institute of Acupuncture Science, China Medical University, Taichung (Taiwan), (2) Department of Pediatrics, Taipei Branch, China Medical University Hospital, Taipei (Taiwan), (3) Graduate Institute of Integrated Medicine, China Medical University, Taichung (Taiwan), (4) Department of Radiology, Taipei Medical University-Wan Fang Hospital, Taipei (Taiwan)

[110] The interpretation of X-ray Computed Microtomography images of rocks as an application of volume image processing and analysis, by Kaczmarczyk J. (1), Dohnalik M. (1), Zalewska J. (1), Cnudde, V. (2) - (1) Oil and Gas Institute, Well Logging Department, Kraków (Poland), (2) The Center for X-Ray Tomography, Department of Geology and Soil Science, Ghent University (Belgium)

[111] Matching 4D Porous Media Fluid Flow GeoPET Data with COMSOL Multiphysics Simulation Results, By Johanna Lippmann-Pipke, Johannes Kulenkampff, Gründig Marion, Michael Richter - Helmholtz-Zentrum Dresden-Rossendorf, Institut of Radiochemistry – Research Site Leipzig, Reactive Transport Division, Leipzig (Germany)

[112] Thermal Characteristics of ThermoBrachytherapy Surface Applicators (TBSA) for Treating Chestwall Recurrence, by K. Arunachalam (1), P. F. Maccarini (1), O. I. Craciunescu (1), J. L. Schlorff (2) and P. R. Stauffer (1) - (1) Department of Radiation Oncology, Duke University, Durham, North Carolina USA, (2) Bionix Development Corporation, Paoli PA USA

[113] The Neanderthal face is not cold adapted - By Todd C. Rae (1), Thomas Koppe (2), Chris B. Stringer (3) - (1) Centre for Research in Evolutionary Anthropology, Department of Life Sciences, Roehampton University, UK, (2) Institut für Anatomie und Zellbiologie, Ernst-Moritz-Arndt Universität Greifswald, Germany, (3) Department of Palaeontology, The Natural History Museum, London, UK

[114] Quantification and Segmentation of Electron Tomography Data – Exemplified at ErSi2 Nanocrystals in SiC, by J. Leschner, A. Chuvilin, J. Biskupek and U. Kaiser - Central Facility of Electron Microscopy, Group of Materials Science, Ulm University

[115] A Three-Dimensional Multiscale Model for Gas Exchange in Fruit, by Quang Tri Ho (1), Pieter Verboven (1), Bert E. Verlinden (1), Els Herremans (1), Martine Wevers (2), Jan Carmeliet (3, 4), and Bart M. Nicolaï (1)* - (1) Flanders Center of Postharvest Technology, BIOSYST-MeBioS, and (2) Research Group of Materials Performance and Nondestructive Evaluation (M.W.), Katholieke Universiteit Leuven (Belgium); (3) Building Physics, Swiss Federal Institute of Technology Zurich (ETHZ) (Switzerland); (4) Laboratory for Building Science and Technology, Swiss Federal Laboratories for Materials Testing and Research (Empa) (Switzerland)

[116] THREE-DIMENSIONAL IMAGING OF PORE STRUCTURES INSIDE LOW-k DIELECTRICS, By, Huolin L. Xin (1), Peter Ercius (2), Kevin J. Hughes (3), James R. Engstrom (3), and David A. Muller (2) - (1) Department of Physics, Cornell University, USA; (2) School of Applied and Engineering Physics, Cornell University, USA; (3) School of Chemical and Biomolecular Engineering, Cornell University, USA

[117] Spitting behaviour and fang morphology of spitting cobras, by RA Berthé - 2011

[118] Radionuclide Retention in Concrete Wasteforms, By CC Bovaird, DM Wellman, DP Jansik, MI Wood - US Department Energy

[119] A new salamander from the late Paleocene–early Eocene of Ukraine, by PAVEL P. SKUTSCHAS (1) AND YURI M. GUBIN (2) - (1) Saint Petersburg State University, Vertebrate Zoology Department, Faculty of Biological and Soil Sciences, Saint Petersburg, Russian Federation; (2) Paleontological Institute, Russian Academy of Sciences, Moscow, Russian Federation.

[120] Beyond the Closed Suture in Apert Syndrome Mouse Models: Evidence of Primary Effects of FGFR2 Signaling on Facial Shape at Birth - By Neus Martinez-Abadias (1), Christopher Percival (1), Kristina Aldridge (2), Cheryl A. Hill (2), Timothy Ryan (1), Satama Sirivunnabood (1), Yingli Wang (3), Ethylin Wang Jabs (3) and Joan T. Richtsmeier (1)* - (1) Department of Anthropology, Pennsylvania State University, University Park, Pennsylvania, (2) Department of Pathology & Anatomical Sciences, University of Missouri-School of Medicine, Columbia, Missouri, (3) Department of Genetics and Genomic Sciences, Mount Sinai School of Medicine, One Gustave L. Levy Place, New York - Grant sponsors: National Institutes of Craniofacial and Dental Research (NIDCR), National Institutes of Health (NIH); Comissionat per a Universitats i Recerca (CUR), Generalitat de Catalunya, Spain.

[121] CHARACTERIZATION OF GEOMORPHIC PORE STRUCTURE OF BUILDING MATERIALS FOR AGENT FATE, by C.R. Savidge, L.B. Hu,N.J. Hayden, D.M. Rizzo, M.M. Dewoolkar - School of Engineering, University of Vermont (USA)

[122] TRANSPORT PHENOMENA ON THE CHANNEL-RIB SCALE OF POLYMER ELECTROLYTE FUEL CELLS, by Reto Flückiger, Dipl. Masch. Ing. ETH Zurich - 2009

[123] ON THE INSULATOR-CONDUCTOR TRANSITION IN POLYMER NANOCOMPOSITES, by Gianluca AMBROSETTI - ÉCOLE POLYTECHNIQUE FÉDÉRALE DE LAUSANNE (Switzerland) - 3D reconstruction of X-ray tomography phase contrast images of a BNB90-polypropylene composite

[124] COMPARISON OF ULTRA-FAST MICROWAVE SINTERING AND CONVENTIONAL THERMAL SINTERING IN MANUFACTURING OF ANODE SUPPORT SOLID OXIDE FUEL CELL, by Zhenjun Jiao(1), Naoki Shikazono(1), Nobuhide Kasagi(2) - (1)Institute of Industrial Science, the University of Tokyo. (2)Department of Mechanical Engineering, The University of Tokyo.

[125] Etude de la microstructure d'électrodes de piles à combustible de type SOFC, by Nicolas Vivet(1) - (1)CEA Le RIPAULT, France.

[126] TRABECULAR BONE STRUCTURE IN THE HUMERAL AND FEMORAL HEADS OF ANTHROPOID PRIMATES, by Timothy M. Ryan(1),(2) and Alan Walker(1)- (1) Department of Anthropology, Pennsylvania State University (USA), (2) Center for Quantitative Imaging, Pennsylvania State University (USA)

[127] Computational Simulation of Breast Compression Based on Segmented Breast and Fibroglandular Tissues on Magnetic Resonance Images By Tzu-Ching Shih (1,2,3), Jeon-Hor Chen (2,3), Dongxu Liu (4), Ke Nie (2), Lizhi Sun (4), Muqing Lin (2), Daniel Chang (2), Orhan Nalcioglu(2), and Min-Ying Su (2) - (1) Department of Biomedical Imaging and Radiological Science, China Medical University, Taichung (Taiwan), (2) Tu and Yuen Center for Functional Onco-Imaging, University of California (USA), (3) Department of Radiology, China Medical University Hospital, Taichung (Taiwan), (4) Department of Civil and Environmental Engineering, University of California, Irvine, California (USA)

[128] Microtomography of Partially Molten Rocks: Three-Dimensional Melt Distribution in Mantle Peridotite, By Wenlu Zhu (1), Glenn A. Gaetani (2), Florian Fusseis (3), Laurent G. J. Montési (1), and Francesco De Carlo (4) - (1) Department of Geology, University of Maryland, USA; (2) Department of Geology and Geophysics, Woods Hole Oceanographic Institution, USA; (3) Western Australian Geothermal Centre of Excellence, The University of Western Australia, Australia; (4) Advanced Photon Source, Argonne National Laboratory, USA.

[129] X-ray in-line phase microtomography for biomedical applications by M. Langer (1,2,3), R. Boistel (4,5), E. Pagot (2,6), P. Cloetens (2), F. Peyrin (1,2) - (1) Université de Lyon-CNRS-Inserm-INSA (France), (2) European Synchrotron Radiation Facility (France), (3) Julius Wolff Institut & Berlin-Brandenburg School for Regenerative Therapies (Germany), (4) IPHEP, CNRS, Université de Poitiers (France), (5) CNPS, CNRS, Université Paris-Sud (France)

[130] 3D ANALYSIS OF THE INTERMEDIATE FILAMENT NETWORK UNSING SEM-TOMOGRAPHY, by M. Sailer(1), S. Lück(2), V. Schmidt(2), M. Beil(3), G. Adler(3) and P. Walther(1) - (1)Electron Microscopy Facility, Ulm University, Germany. (2)Institute of Stochastics, Ulm University, Germany. (3)Department of Internal Medicine I, University Hospital Ulm, Germany.

[131] PORE NETWORK MODELLING ON CARBONATE: A COMPARATIVE STUDY OF DIFFERENT MICRO-CT NETWORK EXTRACTION METHODS, by Hu Dong, Ståle Fjeldstad, Luc Alberts, Sven Roth, Stig Bakke and Pål-Eric Øren - Numerical Rocks AS, Norway

[132] CHARACTERIZATION OF POROUS BUILDING MATERIALS FOR AGENT TRANSPORT PREDICTIONS USING ARTIFICIAL NEURAL NETWORKS, by Cabot R. Savidge to The Faculty of the Graduate College of The University of Vermont

[133] A comparative study of X-ray tomographic microscopy on shales at different synchrotron facilities: ALS, APS and SLS, by W. Kanitpanyacharoen, D. Y. Parkinson, F. De Carlo, F. Marone, M. Stampanoni, R. Mokso, A. MacDowell and H.-R. Wenk

[134] Closing the lifecycle for the sustainable aquaculture of the bath sponge Coscinoderma matthewsi, by M.A. Abdul Wahab (a), R. de Nys (a), S. Whalan (a) - (a) School of Marine & Tropical Biology, James Cook University, Townsville, Queensland (Australia)

[135] Microfluidic 3D bone tissue model for high-throughput evaluation of wound-healing and infection-preventing biomaterials, by Joung-Hyun Lee (a), Yexin Gua, Hongjun Wang (b), Woo Y. Lee (a) - (a) Chemical Engineering and Materials Science, Stevens Institute of Technology, Hoboken (USA), (b) Chemistry, Chemical Biology, and Biomedical Engineering, Stevens Institute of Technology, Hoboken (USA)

[136] Root aeration via aerenchymatous phellem: three-dimensional micro-imaging and radial O_2 profiles in Melilotus siculus, By Pieter Verboven,Ole Pedersen,Els Herremans,Quang Tri Ho,Bart M. Nicolaï,Timothy David Colmer,Natasha Teakle - New Phytologist, 2011

[137] Enhancing Appalachian Coalbed Methane Extraction by Microwave-Induced Fractures, by Jonathan P. Mathews - The Pennsylvania State University

[138] FRACTURING CONTROLLED PRIMARY MIGRATION OF HYDROCARBONS FLUIDS DURING HEATING OF ORGANIC-RICH SHALES, by Maya Kobchenko (1), Hamed Panahi (1)(2), François Renard (1)(3), Dag K. Dysthe (1), Anders Malthe-Sørenssen (1), Adriano Mazzini (1), Julien Scheibert (1), Bjørn Jamtveit (1) and Paul Meakin (1) (4) (5) - (1) Physics of Geological Processes, University of Oslo, Norway; (2) Statoil ASA, Norway; (3) Institut des Sciences de la Terre, Université Joseph Fourier-CNRS, Grenoble, France; (4) Idaho National Laboratory, Idaho Falls, USA; (5) Institute for Energy Technology, Kjeller, Norway

[139] VISUALIZATION AND QUANTIFICATION OF BIOFILM ARCHITECTURE WITHIN POROUS MEDIA USING SYNCHROTRON BASED X-RAY COMPUTED MICROTOMOGRAPHY, by Gabriel Iltis, Ryan Armstrong, and Dorthe Wildenschild - Dpt of Chemical, Biological and Environmental Engineering, Oregon State University, USA

[140] Luo, L., et al. QUANTIFICATION OF 3-D SOIL MACROPORE NETWORKS IN DIFFERENT SOIL TYPES AND LAND USES USING COMPUTED MICROTOMOGRAPHY. J. Hydrol. (2010), doi:10.1016/j.jhydrol.2010.03.031

[141] THE IMPACT OF THERMAL AND CHEMICAL EFFECTS IN FRACTURE DEFORMATION, by Ruqayia Al-Zadjalia (School of Earth and Environment, University of Leeds, UK), Suleiman Al-Hinaib (Petroleum Development of Oman, MAF, Sultanate of Oman), Carlos Grattonic (Rock Deformation Research, University of Leeds, UK), and Quentin Fishera (School of Earth and Environment, University of Leeds, UK).

[142] EXPERIMENTAL OBSERVATIONS AND SIMULATIONS OF THE MECHANICAL DEFORMATION OF AMORPHOUS METALLIC FOAM, by Sarah Shiley Haubrich (Iowa State University, USA)

[143] Three-Dimensional Morphological Measurements of LiCoO2 and LiCoO2/Li(Ni1/3Mn1/3Co1/3)O2 Lithium-ion Battery Cathodes, by Zhao Liu(a), J. Scott Cronin(a), Yu-chen K. Chen-Wiegart(b), James. R. Wilson(a), Kyle J. Yakal-Kremski(a), Jun Wang(b), Katherine T. Faber(a), Scott A. Barnett(a) - (a) Department of Materials Science and Engineering, Northwestern University (USA), (b) Photon Science Directorate, Brookhaven National Laboratory (USA)

[144] Wormhole Propagation in Tar During Matrix Acidizing of Carbonate Formations, by S.H. Al-Mutairi, SPE, M.A. Al-Obied, SPE, I.S. Al-Yami, SPE, A.M. Shebatalhamd, SPE, D.A. Al-Shehri, Saudi Aramco

[145] The Dynamics of Embolism Repair in Xylem: *In Vivo* Visualizations Using High-Resolution Computed Tomography, by Craig R. Brodersen (1), Andrew J. McElrone (1)(3), Brendan Choat (4), Mark A. Matthews (1), and Kenneth A. Shackel (2) - (1) Department of Viticulture and Enology and (2) Department of Plant Sciences, University of California, USA; (3) Department of Agriculture-Agricultural Research Service, Crops Pathology and Genetics Research Unit, Davis, California, USA; (4) Research School of Biology, Australian National University, Canberra, Australia

[146] MEASUREMENT AND PREDICTION OF THE RELATIONSHIP BETWEEN CAPILLARY PRESSURE, SATURA-TION, AND INTERFACIAL AREA IN A NAPL-WATER-GLASS BEAD SYSTEM (2010), by Mark L. Porter (1), Dorthe Wildenschild (1), Gavin Grant (2), and Jason I. Gerhard (3) - (1)School of Chemical, Biological, and Environmental Engineering,Oregon State University, USA. (2)Geosyntec Consultants, Guelph, Ontario, Canada. (3)Department of Civil and Environmental Engineering, University of Western Ontario, Canada.

[147] Huolin L. Xin, Peter Ercius, Kevin J. Hughes, James R. Engstrom, and David A. Muller, Appl. Phys. Lett. 96, 223108 (2010)

[148] MICROSTRUCTURE DE MATÉRIAUX ENTRANT DANS LA COMPOSITION DE PILES A COMBUSTIBLE DE TYPE SOFC by Nicolas Vivet, CEA, France (2010)

[149] LOKALISERING VAN MINERALISEERBARE KOOLSTOFPOOLS IN THE BODEMMATRIX AAN DE HAND VAN X-STRALEN TOMOGRAFIE, by Bram Hantson - Howest – dDpartement Academische Bachelor- en Masteropleidingen, Belgium

[150] SIMULATIONS ET ANALYSES 2D-3D DE LA POROSITE DE DEPOTS PLASMA D'ALUMINE POUR LA CAR-ACTERISATION DE PROPRIETES MECANIQUES ET PHYSIQUES by Vincent Guipont, Centre des Matériaux, CNRS (France)

[151] Identification des propriétés des tissus mous de la jambe sous compression élastique, by L. Dubuis (1), S. Avril (1), P. Badel (1), J. Debayle (2) - (1) LCG, École des Mines de Saint-Étienne (France), (2) LPMG, École des Mines de Saint-Étienne (France)

[152] Microtomographic study and finite element analysis of the porosity harmfulness in a cast aluminium alloy, by N. Vanderesse (a), É. Maire (a), A. Chabod (b), J.-Y. Buffière (a) - (a) Université de Lyon, INSA-Lyon, MATEIS CNRS, Villeurbanne (France), (b) Centre Technique des Industries de la Fonderie (CTIF), Sèvres (France)

[153] Transport Phenomena on the Channel-Rib Scale of Polymer Electrolyte Fuel Cells, by Reto Fluckiger - ETH Zurich

[154] ONE-MONTH SIMULATED PLANET SIMULATOR CIRCULATION, Meteorologisches Institut, Universität Hamburg (Germany)

[155] VISUALISATION IMMERSIVE ET INTERACTION HAPTIQUE : UNE REVOLUTION POUR LES GEOSCIENCES, by BRGM (Bureau des recherches géologiques et minières, France)

Chapter 7

AVM Navigator

AVM Navigator is an additional module of the RoboRealm (plugin) that provides object recognition and autonomous robot navigation using a single video camera on the robot as the main sensor for navigation.

7.1 Associative Video Memory

It is possible due to using of an "Associative Video Memory" (AVM) algorithm based on multilevel decomposition of recognition matrices. It provides image recognition with low False Acceptance Rate (about 0.01%). In this case visual navigation is just the sequence of images (landmarks) with associated coordinates that was memorized inside AVM tree during route training. The navigation map is presented as the set of data (such as X, Y coordinates and azimuth) associated with images inside AVM tree. When a robot sees images from camera (marks) that can be recognized then it confirms its current location.

The navigator creates a way from the current location to target position as a chain of waypoints. If the robot's current orientation does not point to the next waypoint then the navigator turns the robot body. When the robot reaches a waypoint the navigator changes direction to the next waypoint in the chain and so on until the target position is reached.

7.2 External links

- Official AVM Navigator help page
- AVM algorithm description and port for Csharp

Chapter 8

Ayotle

Ayotle is a French-Mexican company headquartered in Paris, France, that develops computer vision software and provides technical services based on motion capture and 3D sensors for interactive applications to the media and entertainment industry. The company was co-founded by José Alonso YBANEZ ZEPEDA and Gisèle BELLIOT on June 2010.[1]

Ayotle's expertise centres on the development and implementation of advanced algorithms for computer vision, from video images in all formats, and to the use of 3D cameras or depth sensors.

Up until now, Ayotle developed two main innovative software solutions. With FaceTracker, using highly advanced technologies in markerless motion capture, Ayotle offers new solutions for computer facial animation. With the AnyTouch project, Ayotle provides a solution that can transform any surface or object into a touch device.

8.1 Etymology

The name **Ayotle** comes from the word *Ayotl*, which means turtle shell in Nahuatl, the original language used by the Aztecs in Mexico.

8.2 Supports

Ayotle is currently supported by Paris Region Lab,[2] ASTIA,[3] Mairie de Paris,[4] Oséo,[5] Scientipôle Initiative,[6] Cap-Digital,[7] Telecom ParisTech[8]

8.3 References

[1] Ayotle Home Page

[2] Paris Region Lab

[3] ASTIA

[4] Mairie de Paris

[5] Oséo

[6] Scientipole Initiative

[7] Cap-Digital

[8] Telecom ParisTech

8.4 External links

- June 3, 2012 by Brian Anthony Hernandez "AnyTouch Turns Any Surface Into a Touchscreen [VIDEO]" Mashable

- By Allie Walker on May 31, 2012 "Turn Anything Into A Touchscreen Simply By Touching It" PSFK

- Spotted by: Katherine Noyes "With 3D camera and depth sensors, any object can be touch-enabled" Springwise

- By Bruce Sterling June 17, 2012 "Augmented Reality: AnyTouch" Wired

- Submitted by Adam Pasulka on Wed, 06 Jun 2012 "AnyTouch" Protein

- BY GARETH SWARTE / 18 JUN 2012 "The marketing potential of AnyTouch" ClockWorkDragon

- PUBLIÉ LE 30 MAI 2012, Gaël Clouzard "AnyTouch et le monde devient tactile…" Influencia

- ParisTech Entrepreneurs

- June 5, 2012 at 9:52 by Bridget "The future of tactile navigation? Adverblog

Chapter 9

Bing Audio

Bing Audio (also known as **Bing Music**)[1] is a music recognition application created by Microsoft which is installed on Windows Phones running version 7.5 and above, including Windows Phone 8. On Windows Phone 8.1, and in regions where the Microsoft Cortana voice assistant is available, Bing Music is integrated with Cortana[2] and the music search history is a part of Cortana's "Notebook". The service is only designed to recognize recorded songs, not live performances or humming. Xbox Music Pass subscribers can immediately add the songs to their playlists.[3] A unique feature compared to similar services is that Bing Audio continuously listens and analyzes music while most other services can only listen for a fixed amount of time. Bing Research developed a *fingerprinting* algorithm to identify songs.[4]

9.1 Availability

9.1.1 As part of Cortana (where supported)

- Australia

- Canada

- France

- Germany

- Italy

- Spain

- United Kingdom

- United States

9.1.2 As part of Bing Mobile

- Argentina

- Austria

- Belgium

- Brazil

- Denmark

- Finland

- Ireland

- Mexico

- Netherlands

- New Zealand

- Norway

- Portugal

- Sweden

- Switzerland

[5]

9.2   See also

- Gracenote's MusicID-Stream

- Play by Yahoo Music

- Shazam

- Sony TrackID

- SoundHound

- Groove Music (known as Zune in versions prior to Windows Phone 8, and as Xbox Music in Windows Phone 8.x)

9.3   References

[1] Stroh, Micheal (3 July 2013). "Bing Audio, Windows Phone's built-in music matching service, rolls out to 14 new countries". *Blogging Windows*.

[2] Súrîl, Surur (7 April 2014). "Joe Belfiore confirms WP8.1 Developer Preview is coming "first part of April" and other titbits". *WMPoweruser*.

[3] Burgess, Brian (19 April 2014). "Use Cortana on Windows Phone 8.1 to Identify Songs". *GroovyPost*.

[4] Stroh, Micheal (8 June 2011). "Q&A: The story behind Music search". *Blogging Windows*.

[5] Wang, Abigail (4 July 2013). "Bing Audio Brings Music Discovery to 14 New Countries". *PCMag*.

Chapter 10

Bing Vision

Bing Vision is an image recognition application created by Microsoft which is installed on Windows Phones running version 7.5 and above, including Windows Phone 8. It is a part of the Bing Mobile suite of services, and on most devices can be accessed using the search button. On Windows Phone 8.1 devices where Microsoft Cortana is available, it is only available through the lenses of the Camera app (as the search button now activates Cortana).[1][2] Bing Vision can scan barcodes, QR codes, Microsoft Tags, books, CDs, and DVDs.[3] Books, CDs, and DVDs are offered through Bing Shopping.[4][5]

10.1   See also

- Nokia Point & Find
- Google Goggles

10.2   References

[1] Suril, Surur (17 April 2014). "How to easily launch Bing Vision on WP 8.1". *WMPoweruser*.

[2] Lonut, Arghire (1 August 2014). "Bing Vision in Windows Phone 8.1 No Longer Scanning Barcodes". *Softpedia*.

[3] Smith, Mat (September 5, 2012). "Windows Phone 8 introduces new Lens apps: Bing Vision, Photosynth and CNN iReport launching from the camera button". *Engadget*.

[4] Blandford, Rafe (30 January 2012). "Bing Vision's Book, CD & DVD recognition expanded to UK". *All About Windows Phone*.

[5] Callaham, John (20 October 2012). "Bing features for Windows Phone 8 explained". *Neowin*.

10.3   External links

- Official website

Chapter 11

CellCognition

CellCognition is a free open-source computational framework for quantitative analysis of high-throughput fluorescence microscopy (time-lapse) images in the field of bioimage informatics and systems microscopy. The CellCognition framework uses image processing, computer vision and machine learning techniques for single-cell tracking and classification of cell morphologies. This enables measurements of temporal progression of cell phases, modeling of cellular dynamics and generation of phenotype map.[1][2]

11.1 Features

CellCognition uses a computational pipeline which includes image segmentation, object detection, feature extraction, statistical classification, tracking of individual cells over time, detection of class-transition motifs (e.g. cells entering mitosis), and HMM correction of classification errors on class labels.

The software is a cross-platform application and runs on the three major operating systems (Microsoft Windows, Mac OS X, and Linux). It combines VIGRA based C++ computer vision library with Python based workflow engine and graphical user interface.

11.2 History

CellCognition (Version 1.0.1) was first released in December 2009 by scientists from the Gerlich Lab and the Buhmann group at the Swiss Federal Institute of Technology Zürich and the Ellenberg Lab at the European Molecular Biology Laboratory Heidelberg. The current version is 1.2.5 and being developed and maintained by the Gerlich Lab at the Institute of Molecular Biotechnology.

11.3 Application

CellCognition has been used in RNAi-based screening,[3] applied in basic cell cycle study,[4] and extended to unsupervised modeling.[5]

11.4 Resources

CellCognition project is hosted on GitHub and allows any interested developer to contribute to the project.

11.5   References

[1] Held M, Schmitz MHA, Fischer B, Walter T, Neumann B, Olma MH, Peter M, Ellenberg J, Gerlich DW (2010). "CellCognition: time-resolved phenotype annotation in high-throughput live cell imaging". *Nature Methods* **7** (9): 747–54. doi:10.1038/nmeth.1486. PMID 20693996.

[2] Schmitz MHA, Held M, Janssens V, Hutchins JR, Hudecz O, Ivanova E, Goris J, Trinkle-Mulcahy L, Lamond AI, Poser I, Hyman AA, Mechtler K, Peters JM, Gerlich DW (2010). "Live-cell imaging RNAi screen identifies PP2A-B55alpha and importin-beta1 as key mitotic exit regulators in human cells". *Nature Cell Biology* **12** (9): 886–93. doi:10.1038/ncb2092. PMID 20711181.

[3] Piwko W, Olma MH, Held M, Bianco JN, Pedrioli PG, Hofmann K, Pasero P, Gerlich DW, Peter M (2010). "RNAi-based screening identifies the Mms22L-Nfkbil2 complex as a novel regulator of DNA replication in human cells". *EMBO Journal* **29** (24): 4210–22. doi:10.1038/emboj.2010.304. PMC 3018799. PMID 21113133.

[4] Wurzenberger C, Held M, Lampson MA, Poser I, Hyman AA, Gerlich DW (2012). "Sds22 and Repo-Man stabilize chromosome segregation by counteracting Aurora B on anaphase kinetochores". *EMBO Journal* **198** (2): 173–83. doi:10.1083/jcb.201112112. PMC 3410419. PMID 22801782.

[5] Zhong Q, Busetto AG, Fededa JP, Buhmann JM, Gerlich DW (2012). "Unsupervised modeling of cell morphology dynamics for time-lapse microscopy". *Nature Methods* **9** (7): 711–13. doi:10.1038/nmeth.2046. PMID 22635062.

11.6   External links

- CellCognition website

- GitHub project page

Chapter 12

CVIPtools

CVIPtools (Computer Vision and Image Processing Tools) is an Open Source image processing software.[1] It is free for use with Windows, and previous versions are available for UNIX. It is an interactive program for image processing and computer vision.[2][3]

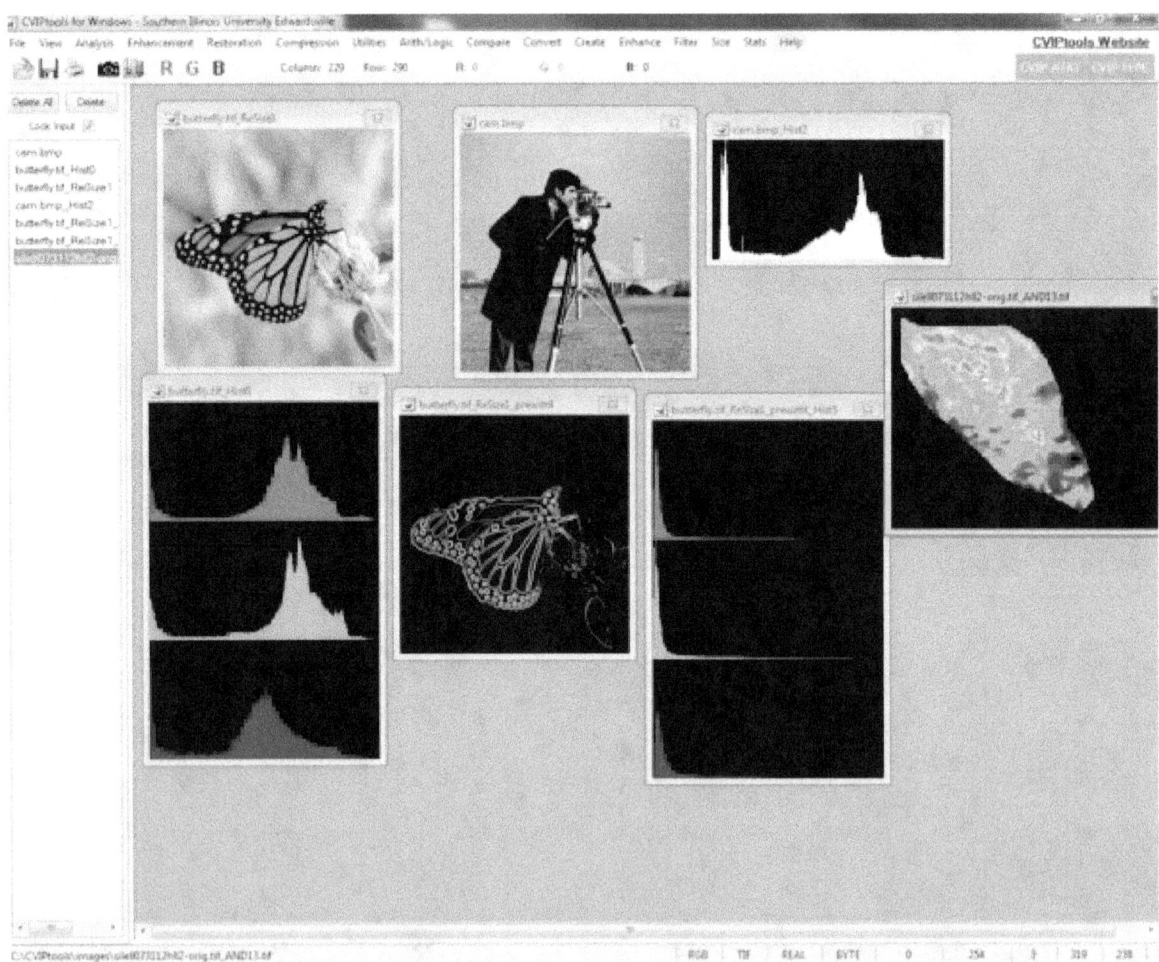

Running the CVIPtools

12.1   About CVIPtools

CVIPtools is a Windows-based software (previous versions for various flavors of UNIX are also available) for computer vision and image processing developed at the Computer Vision and Image Processing Laboratory at Southern Illinois University at Edwardsville.[4] CVIPtools 5.5d is implemented in four layers: the algorithms code layer, the Common Object Module (COM) interface layer, the CvipOp layer, and the Graphical user interface (GUI). The algorithms code layer is based primarily on previous versions of CVIPtools, consists of all image and data-processing procedures and functions, and is written in standard C. The COM interface layer is written in C++ and links the CVIPtools C functions to the GUI through the CVIPimage Class. The CvipOp layer provides an object-oriented paradigm by using the Class CVIPimage to consolidate data safety and memory management. The GUI layer, written in C#, implements the image queue, viewer, and manages user input and output. For development, CVIPtools5.x includes the CVIPlab environment and the CVIPtools libraries. In addition to the standard C libraries, a dynamically linked library (cviptools.dll) is provided that contains all the COM versions of these functions.

12.2   Features

CVIPtools can read many image formats including TIFF, PNG, GIF, JPEG, BMP, as well as raw formats. CVIPtools supports standard image processing functions such as image compression, image restoration, logical and arithmetical operations between images, contrast manipulation, image sharpening, Frequency transform, edge detection, segmentation and geometric transformations.[5]

CVIPtools5.x also contains two powerful development tools that allow for batch processing and automatic algorithm analysis and development. The CVIP-ATAT, Algorithm Test and Analysis Tool, can be used to test all combinations and values of parameters to speed front-end algorithm development. The CVIP-FEPC, Feature Extraction and Pattern Classification, will allow for batch processing and test all combinations of features and pattern classification techniques. These are described in more detail, along with application examples, in the new edition of the textbook Digital Image Processing and Analysis: Human and Computer Vision Applications with CVIPtools, Second Edition.[6][7]

12.3   CVIP-ATAT

The Computer Vision and Image Processing Algorithm Test and Analysis Tool, CVIP-ATAT, was created to facilitate the development of both human and computer vision applications. The primary function of this tool is to allow the user to explore many more algorithmic possibilities than can be considered by processing one image at a time with CVIPtools. It allows for the automatic processing of large image sets with many different algorithmic and parameter variations. We call this the "front-end" tool because its primary purpose is to find the best algorithm to preprocess, segment and post-process a set of images for a particular application in order to best separate the most important regions of interest within the image.

It has a GUI which allows the user to enter multi-stage algorithms for testing and analysis. At each stage the user can specify a number of different processes to test and a range for the processes' parameters. The user also specifies a set of images to process and a set of "ideal" output images which will be used to determine the success for each algorithm. Note that one algorithm is defined as a specific set of processes and a specific set of parameter values.

The tool will then automatically perform algorithms which consist of all the permutations of the values for each of the parameters for each process and all the processes for each stage. Next, the user can compare the various algorithm results to determine the best set of processes and parameters for the particular application. The tool is useful for application development where the ideal image results are available, or can be created. Additionally, it can serve as a front end development tool for image analysis to find the optimal set of processes and parameters for extracting regions of interest for further processing.

12.4 CVIP-FEPC

The Computer Vision and Image Processing Feature Extraction and Pattern Classification Tool, CVIP-FEPC, was created to facilitate the development of both human and computer vision applications. The primary application area is computer vision, but it can be used, for example, as an aid in the development of image compression schemes for human vision applications. This can be done by helping to determine salient image features that must be retained for a given compression scheme. Conversely, computer vision applications are essentially deployed image analysis systems for a specific application, so the feature extraction and pattern classification is an integral part of all computer vision systems.

The primary function of this tool is to explore feature extraction and pattern classification and allow the user to perform batch processing with large image sets and is thus much more efficient than processing one image at a time with CVIPtools. It allows the user to select the features and pattern classification parameters for the automatic processing of these large image sets. CVIP-FEPC enables the user to easily specify the training and test sets and run multiple experiments in an efficient manner. Its primary purpose is to find the best parameters for a particular application in order to best classify the image objects of interest.

This tool is designed to work with a set of images that have binary masks that have been created for the objects of interest – one object per image. These masks can be created manually with CVIPtools, or, many image database applications will have the masks available. In general, the user will load the images, specify the classes, select the features, select the test set, choose the pattern classification parameters and then let the program process the entire image set. An output file will be created with the results for the experiment.

12.5 References

[1] Image Processing Software: http://www.imageprocessingplace.com/root_files_V3/software/software.htm

[2] http://www.imageprocessingplace.com/root_files_V3/software/software.htm

[3] http://cviptools.software.informer.com/

[4] CVIPtools Developer Site: http://cviptools.ece.siue.edu

[5] http://fileforum.betanews.com/detail/CVIPtools-for-Linux/1017471208/3

[6] Umbaugh, Scott E (2010). *Digital image processing and analysis : human and computer vision applications with CVIPtools* (2nd ed.). Boca Raton, FL: CRC Press. ISBN 9-7814-3980-2052.

[7] http://www.crcpress.com/product/isbn/9781439802052

Chapter 13

DeepDream

An image of a toast sandwich passed through the DeepDream *program.*

DeepDream is a computer vision program created by Google which uses a convolutional neural network to find and enhance patterns in images via algorithmic pareidolia, thus creating a dreamlike hallucinogenic appearance in the deliberately over-processed images.[1][2][3]

13.1   Software

The DeepDream software, initially codenamed Inception after the film *Inception*,[3][1][2] was developed for the ImageNet Large-Scale Visual Recognition Challenge (ILSVRC) in 2014[3] and released in July 2015. The software is designed to detect faces and other patterns in images, with the aim of automatically classifying images.[4] (The software also attempts to ascertain what is happening in a picture, making attempts to form sentences which describe it.)[5]

When reiterations are run to tease out the found imagery even further, a form of pareidolia results by which psychedelic and surreal images are generated algorithmically. The oft-cited resemblance of the imagery to LSD and psilocybin induced hallucinations is suggestive of a functional resemblance between artificial neural networks and particular layers of the visual cortex, a matter which merits further study.[6]

After Google made the code open source,[7] a number of websites to run it were founded which enabled users to subject their own digital photos to the program.[8] Desktop apps have also been established like Deep Dreamer for OS X, a paid application for Mac that requires Yosemite. The DeepDream framework has also become available in the form of mobile apps, offering specific Deep Dream photo-editing filters, such as the free iOS and Android app Dreamception. [9]

13.2 References

[1] Mordvintsev, Alexander; Olah, Christopher; Tyka, Mike (2015). "DeepDream - a code example for visualizing Neural Networks". Google Research. Archived from the original on 2015-07-08.

[2] Mordvintsev, Alexander; Olah, Christopher; Tyka, Mike (2015). "Inceptionism: Going Deeper into Neural Networks". Google Research. Archived from the original on 2015-07-03.

[3] Szegedy, Christian; Liu, Wei; Jia, Yangqing; Sermanet, Pierre; Reed, Scott; Anguelov, Dragomir; Erhan, Dumitru; Vanhoucke, Vincent; Rabinovich, Andrew (2014). "Going Deeper with Convolutions" (PDF). *Computing Research Repository*. arXiv:1409.4842.

[4] Rich McCormick (7 July 2015). "Fear and Loathing in Las Vegas is terrifying through the eyes of a computer". *The Verge*. Retrieved 2015-07-25.

[5] Rich McCormick (2015-07-17). "First computers recognized our faces, now they know what we're doing". *The Verge*. Retrieved 2015-07-25.

[6] LaFrance, Adrienne. "When Robots Hallucinate". The Atlantic. Retrieved 24 September 2015.

[7] deepdream on GitHub

[8] Daniel Culpan (2015-07-03). "These Google "Deep Dream" Images Are Weirdly Mesmerising". *Wired*. Retrieved 2015-07-25.

[9] Mark Gibbs (2015-08-24). "The ever-expanding Deep Dreaming alternate universe gets an iOS app". *NetworkWorld*. Retrieved 2015-08-25.

13.3 External links

- Deep Dream, Ipython notebook at GitHub

- Mordvintsev, Alexander; Olah, Christopher; Tyka, Mike (June 17, 2015). "Inceptionism: Going Deeper into Neural Networks". Google Research.

- Deep Dream Generator

- Dreamception - iOS & Android DeepDream App

Chapter 14

Dlib

Not to be confused with Digital Library of Slovenia.

Dlib is a general purpose cross platform open source software library written in the C++ programming language. Its design is heavily influenced by ideas from design by contract and component-based software engineering. This means it is, first and foremost, a collection of independent software components.

Since development began in 2002, dlib has grown to include a wide variety of tools. In particular, it now contains software components for dealing with networking, threads, graphical user interfaces, data structures, linear algebra, machine learning, image processing, data mining, XML and text parsing, numerical optimization, Bayesian networks, and numerous other tasks. In recent years, much of the development has been focused on creating a broad set of statistical machine learning tools and in 2009 dlib was published in the *Journal of Machine Learning Research*.[2] Since then it has been used in a wide range of domains.[3][4][5][6][7][8][9][10][11][12][13][14][15]

14.1 References

[1] "Release notes". *dlib C++ Library.*

[2] King, D. E. (2009). "Dlib-ml: A Machine Learning Toolkit" (PDF). *J. Mach. Learn. Res.* **10** (Jul): 1755–1758. CiteSeerX: 10.1.1.156.3584.

[3] Scholarly research using dlib

[4] dlib on mloss.org

[5] Autonome Mobile Systeme 2009

[6] ESS: Extremely Simple Serialization for C++

[7] Gould, S. (2012). "DARWIN: A Framework for Machine Learning and Computer Vision Research and Development" (PDF). *J. Mach. Learn. Res.* **13** (Dec): 3533–3537. CiteSeerX: 10.1.1.413.8518.

[8] Yan, Junchi, et al. "Online incremental regression for electricity price prediction." Service Operations and Logistics, and Informatics (SOLI), 2012 IEEE International Conference on. IEEE, 2012. Yan, J.; Tian, C.; Wang, Y.; Huang, J. (2012). "Online incremental regression for electricity price prediction". *Proceedings of 2012 IEEE International Conference on Service Operations and Logistics, and Informatics.* p. 31. doi:10.1109/SOLI.2012.6273500. ISBN 978-1-4673-2401-4.

[9] Kuijf, Hugo J., Max A. Viergever, and Koen L. Vincken. "Automatic Extraction of the Curved Midsagittal Brain Surface on MR Images." Medical Computer Vision. Recognition Techniques and Applications in Medical Imaging. Springer Berlin Heidelberg, 2013. 225-232. Kuijf, H. J.; Viergever, M. A.; Vincken, K. L. (2013). "Automatic Extraction of the Curved Midsagittal Brain Surface on MR Images". *Medical Computer Vision. Recognition Techniques and Applications in Medical Imaging.* Lecture Notes in Computer Science **7766**. p. 225. doi:10.1007/978-3-642-36620-8_22. ISBN 978-3-642-36619-2.

[10] Bormann, Richard Klaus Eduard. "Vision-based place categorization." (2010).

[11] Brodu, Nicolas, and Dimitri Lague. "3D terrestrial lidar data classification of complex natural scenes using a multi-scale dimensionality criterion: Applications in geomorphology." ISPRS Journal of Photogrammetry and Remote Sensing 68 (2012): 121–134.

[12] Aung, Zeyar, et al. "Towards accurate electricity load forecasting in smart grids." DBKDA 2012, The Fourth International Conference on Advances in Databases, Knowledge, and Data Applications. 2012.

[13] Rodriguez, Alberto, et al. "Abort and retry in grasping." Intelligent Robots and Systems (IROS), 2011 IEEE/RSJ International Conference on. IEEE, 2011. Rodriguez, A.; Mason, M. T.; Srinivasa, S. S.; Bernstein, M.; Zirbel, A. (2011). "Abort and retry in grasping". *2011 IEEE/RSJ International Conference on Intelligent Robots and Systems*. p. 1804. doi:10.1109/IROS.2011.6095100. ISBN 978-1-61284-456-5.

[14] Mohan, Vandana, et al. "Intraoperative prediction of tumor cell concentration from Mass Spectrometry Imaging." Int. Symp. Math. Theo. Netw. Syst. 2010.

[15] Nakashima, Yuta, Noboru Babaguchi, and Jianping Fan. "Detecting intended human objects in human-captured videos." Computer Vision and Pattern Recognition Workshops (CVPRW), 2010 IEEE Computer Society Conference on. IEEE, 2010. Nakashima, Y.; Babaguchi, N.; Fan, J. (2010). "Detecting intended human objects in human-captured videos". *2010 IEEE Computer Society Conference on Computer Vision and Pattern Recognition - Workshops*. p. 33. doi:10.1109/CVPRW.2010.5543721. ISBN 978-1-4244-7029-7.

14.2 External links

- Official website

- dlib C++ Library on SourceForge.net

- DLib: Library for Machine Learning

Chapter 15

Fiji (software)

Fiji (*Fiji Is Just ImageJ*)[1][2] is an open source image processing package based on ImageJ.

Fiji's main purpose is to provide a distribution of ImageJ with many bundled plugins. Fiji features an integrated updating system and aims to provide users with a coherent menu structure, extensive documentation in the form of detailed algorithm descriptions and tutorials, and the ability to avoid the need to install multiple components from different sources.

Fiji is also targeted at developers, through the use of a version control system, an issue tracker, dedicated development channels and a rapid-prototyping infrastructure in the form of a script editor which supports BeanShell, Jython, JRuby and other scripting languages, as well as Just-In-Time Java development.

15.1   Plugins

Many plugins exist for ImageJ, with a wide range of applications, but also a wide range of quality.[3]

Further, some plugins require specific versions of ImageJ, specific versions of third-party libraries, or additional Java components such as the Java compiler or Java3D.

One of Fiji's principal aims is to make the installation of ImageJ, Java, Java 3D, the plugins, and further convenient components, as easy as possible. As a consequence, Fiji enjoys more and more active users.[4]

15.2   Audience

While Fiji was originally intended for neuro-scientists (and continues to be so[5]), it accumulated enough functionality to attract scientists from a variety of fields, such as cell biology,[6] parasitology,[7] genetics, life sciences in general, material science, etc. As stated on the official website, the primary focus is "life sciences", although Fiji provides many tools helping with scientific image analysis in general.[8]

Fiji is most popular in the Life sciences community, where the 3D Viewer[9] helps visualizing data obtained through light microscopy, and for which Fiji provides registration,[10] segmentation and other advanced image processing algorithms.

The Fiji component TrakEM2 was successfully used and enhanced to analyze neuronal lineages in larval *Drosophila* brains.[11]

Fiji was prominently featured in Nature Methods review supplement on visualization [12]

15.3   Development

Fiji is fully Open Source; its sources live in a Git repository (see the homepage for details).

Fiji was accepted as organization into the Google Summer of Code 2009, and completed two projects.

The scripting framework, which supports Javascript, Jython, JRuby, Clojure, BeanShell and other languages, is an integral part of the development of Fiji; many developers prototype their plugins in one of the mentioned scripting languages, and gradually turn the prototypes into proper Java code. To this end, as one of the aforementioned Google Summer of Code projects, a script editor was added with syntax highlighting and in-place code execution.

The scripting framework is included in the Fiji releases, so that advanced users can use such scripts in their common workflow.

The development benefits from occasional hackathons, where life scientists with computational background meet and improve their respective plugins of interest.

15.4   Script Editor

The script editor in Fiji supports rapid prototyping of scripts and ImageJ plugins, making Fiji a powerful tool to develop new image processing algorithms and explore new image processing techniques with ImageJ.[13][14]

15.5   Supported platforms

Fiji runs on Windows, Linux and MacOSX, Intel 32-bit or 64-bit, with limited support for MacOSX/PPC.

15.6   References

[1] Primary reference: Johannes Schindelin, Ignacio Arganda-Carreras, Erwin Frise, Verena Kaynig, Mark Longair, Tobias Pietzsch, Stephan Preibisch, Curtis Rueden, Stephan Saalfeld, Benjamin Schmid, Jean-Yves Tinevez, Daniel James White, Volker Hartenstein, Kevin Eliceiri, Pavel Tomancak and Albert Cardona (2012). "Fiji: an open-source platform for biological-image analysis". *Nature Methods* **9(7)** (7): 676–682. doi:10.1038/nmeth.2019.

[2] Fiji was presented publicly for the first time on the ImageJ User and Developer Conference in November 2008.

[3] Compare the presentations at the 2nd ImageJ User and Developer Conference in November 2008 and the 3rd ImageJ and User Developer Conference in October 2010.

[4] Compare with the Fiji Usage Map

[5] Longair Mark, Baker DA, Armstrong JD. (2011). "Simple Neurite Tracer: Open Source software for reconstruction, visualization and analysis of neuronal processes". *Bioinformatics* **27** (17): 2453–4. doi:10.1093/bioinformatics/btr390. PMID 21727141.

[6] Preibisch S and Saalfeld S and Tomancak P (April 2009). "Globally Optimal Stitching of Tiled 3D Microscopic Image Acquisitions". *Bioinformatics* **25** (11): 1463–5. doi:10.1093/bioinformatics/btp184. PMC 2682522. PMID 19346324.

[7] Hegge S and Kudryashev M and Smith A and Frischknecht F (May 2009). "Automated classification of *Plasmodium* sporozoite movement patterns reveals a shift towards productive motility during salivary gland infection". *Biotechnology Journal* **4** (6): 903–13. doi:10.1002/biot.200900007. PMID 19455538.

[8] The Fiji Wiki, accessed 2012-11-01.

[9] Benjamin Schmid , Johannes Schindelin , Albert Cardona , Mark Longair and Martin Heisenberg (2010). "A high-level 3D visualization API for Java and ImageJ". *BMC Bioinformatics* **11**: 274. doi:10.1186/1471-2105-11-274. PMC 2896381. PMID 20492697.

[10] Stephan Preibisch, Stephan Saalfeld, Johannes Schindelin and Pavel Tomancak (2010). "Software for bead-based registration of selective plane illumination microscopy data". *Nature Methods* **7** (6): 418–419. doi:10.1038/nmeth0610-418. PMID 20508634.

[11] Albert Cardona, Stephan Saalfeld, Ignacio Arganda, Wayne Pereanu, Johannes Schindelin and Volker Hartenstein (2010). "Identifying Neuronal Lineages of Drosophila by Sequence Analysis of Axon Tracts". *The Journal of Neuroscience* **30** (22): 7538–7553. doi:10.1523/JNEUROSCI.0186-10.2010. PMC 2905806. PMID 20519528.

[12] Thomas Walter, David W Shattuck, Richard Baldock, Mark E Bastin, Anne E Carpenter, Suzanne Duce, Jan Ellenberg, Adam Fraser, Nicholas Hamilton, Steve Pieper, Mark A Ragan, Jurgen E Schneider, Pavel Tomancak and Jean-Karim Hériché (2010). "Visualization of image data from cells to organisms". *Nature Methods* **7** (3s): S26–S41. doi:10.1038/nmeth.1431.

[13] *Scripting in Fiji (Fiji Is Just ImageJ)* at 3rd User and Developer Conference in October 2010

[14] Albert Cardona's crash course *Jython scripting with Fiji*.

15.7   External links

- Official website

- ImageJ2, a closely related project that will provide the new central piece (fully backwards-compatible to ImageJ) of Fiji in the near future.

Chapter 16

GemIdent

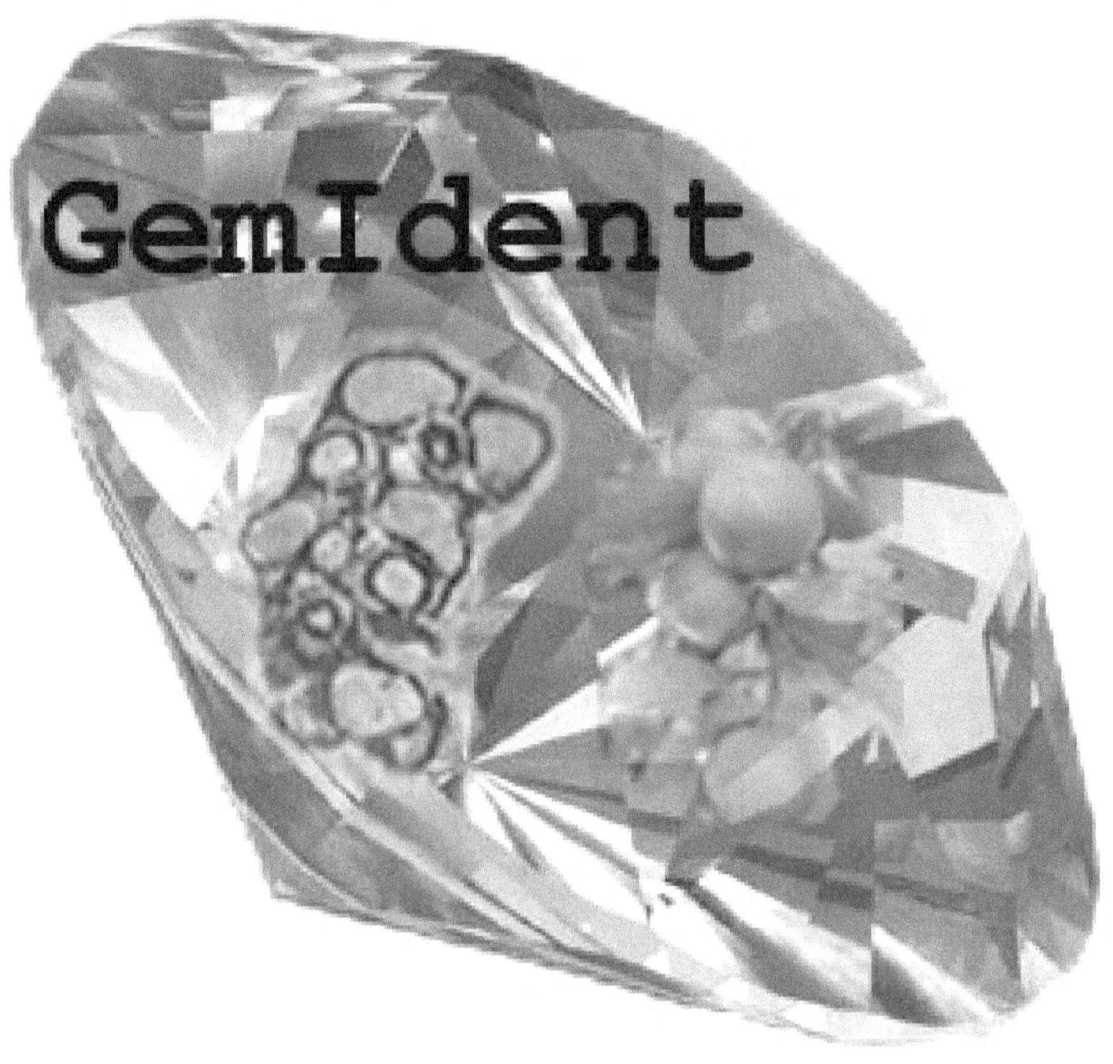

GemIdent logo

GemIdent is an interactive image recognition program that identifies regions of interest in images and photographs. It is specifically designed for images with few colors, where the objects of interest look alike with small variation. For example, color image segmentation of:

- Oranges from a tree

- Stained cells from microscopic images

GemIdent also packages data analysis tools to investigate spatial relationships among the objects identified.

16.1 History

GemIdent was developed at Stanford University by Adam Kapelner from June, 2006 until January, 2007 in the lab of Dr. Peter Lee under the tutelage of Professor Susan Holmes.[1] The concept was inspired by data Kohrt et al.[2] who analyzed immune profiles of lymph nodes in breast cancer patients. Hence, GemIdent works well when identifying cells in IHC-stained tissue imaged via automated light microscopy when the nuclear background stain and membrane/cytoplasmic stain are well-defined. In 2008, it was adapted to support multispectral imaging techniques.[3] It has also recently (July, 2009) been extended to support outsourcing the training to Amazon's MTurk[4] using the extension called "DistributeEyes"[5]

16.2 Methodology

GemIdent uses supervised learning to perform automated identification of regions of interest in the images. Therefore, the user must do a substantial amount of work first supplying the relevant colors, then pointing out examples of the objects or regions themselves as well as negatives (training set creation).

When a user clicks on a pixel, many scores are generated using the surrounding color information via *Mahalanobis Ring Score* attribute generation (read the JSS paper for a detailed exposition). These scores are then used to build a random forest machine-learning classifier which will then classify pixels in any given image.

After classification, there may be mistakes. The user can return to training and point out the specific mistakes and then reclassify. These training-classifying-retraining-reclassifying iterations (considered interactive boosting) can result in a highly accurate segmentation.

16.3 Recent Applications

In 2010, Setiadi et al.[6] analyzed histological sections of lymph nodes looking at spatial densities of B and T cells. "Cell numbers do not capture the full range of information encoded within tissues".

16.4 Source code

The Java source code is now open source under GPL2.[7]

16.5 Examples

The raw photograph (left), a superimposed mask showing the pixel classification results (center), and finally the photograph is marked with the centroids of the object of interest - the oranges (right)

The raw microscopic image of a stained lymph node (left) from the Kohrt study,[2] a superimposed mask showing the pixel classification results (center), and finally the image is marked with the centroids of the object of interest - the cancer nuclei (right)

GemIdent identifying oranges in an orange grove

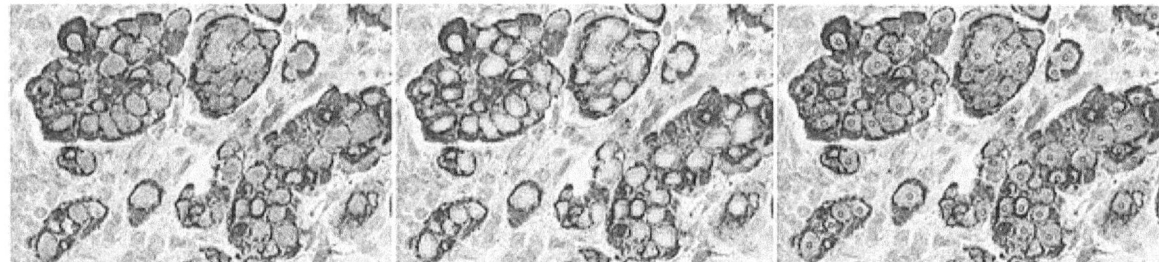

GemIdent identifying cancer cells in a microscopic image

This example illustrates GemIdent's ability to find multiple phenotypes in the same image: the raw microscopic image of a stained lymph node (top left) from the Kohrt study,[2] a superimposed mask showing the pixel classification results (top right), and finally the image marked with the centroids of the objects of interest - the cancer nuclei (in green stars), the T-cells (in yellow stars), and non-specific background nuclei (in cyan stars).

The command-line data analysis and visualization interface in action analyzing results of a classification of a lymph node from the Kohrt study.[2] The histogram displays the distribution of distances from T-cells to neighboring cancer cells. The binary image of cancer membrane is the result of a pixel-only classification. The open PDF document is the autogenerated report of the analysis which includes a thumbnail view of the entire lymph node, counts and Type I error rates for all phenotypes, as well as a transcript of the analyses performed.

16.6   References

[1] Kapelner, Adam; Peter P. Lee; Susan Holmes (July 2007). "An Interactive Statistical Image Segmentation and Visualization System". *Medivis* (IEEE Computer Society) **0**: 81–86. doi:10.1109/MEDIVIS.2007.5. ISBN 0-7695-2904-6.

[2] Kohrt, Holbrook E; Navid Nouri; Kent Nowels; Denise Johnson; Susan Holmes; Peter P Lee (September 2005). "Profile of Immune Cells in Axillary Lymph Nodes Predicts Disease-Free Survival in Breast Cancer". *PLoS medicine* **2** (9): e284. doi:10.1371/journal.pmed.0020284. ISSN 1549-1676. PMC 1198041. PMID 16124834.

[3] Holmes, Susan; Adam Kapelner; Peter P. Lee (January 15, 2009). "An Interactive Java Statistical Image Segmentation System: GemIdent". *Journal of Statistical Software* **30** (10): 1–20. ISSN 1548-7660.

[4] http://mturk.com

[5] http://distributeeyes.com

[6] Setiadi, Francesca; Nelson C. Ray, Holbrook E. Kohrt, Adam Kapelner, Valeria Carcamo-Cavazos, Edina B. Levic, Sina Yadegarynia, Chris M. van der Loos, Erich J. Schwartz, Susan Holmes, Peter P. Lee (Aug 25, 2010). "Quantitative, Architectural Analysis of Immune Cell Subsets in Tumor-Draining Lymph Nodes from Breast Cancer Patients and Healthy Lymph Nodes". *PLoS ONE* **5** (8): 1–20. doi:10.1371/journal.pone.0012420.

[7] https://github.com/kapelner/GemIdent

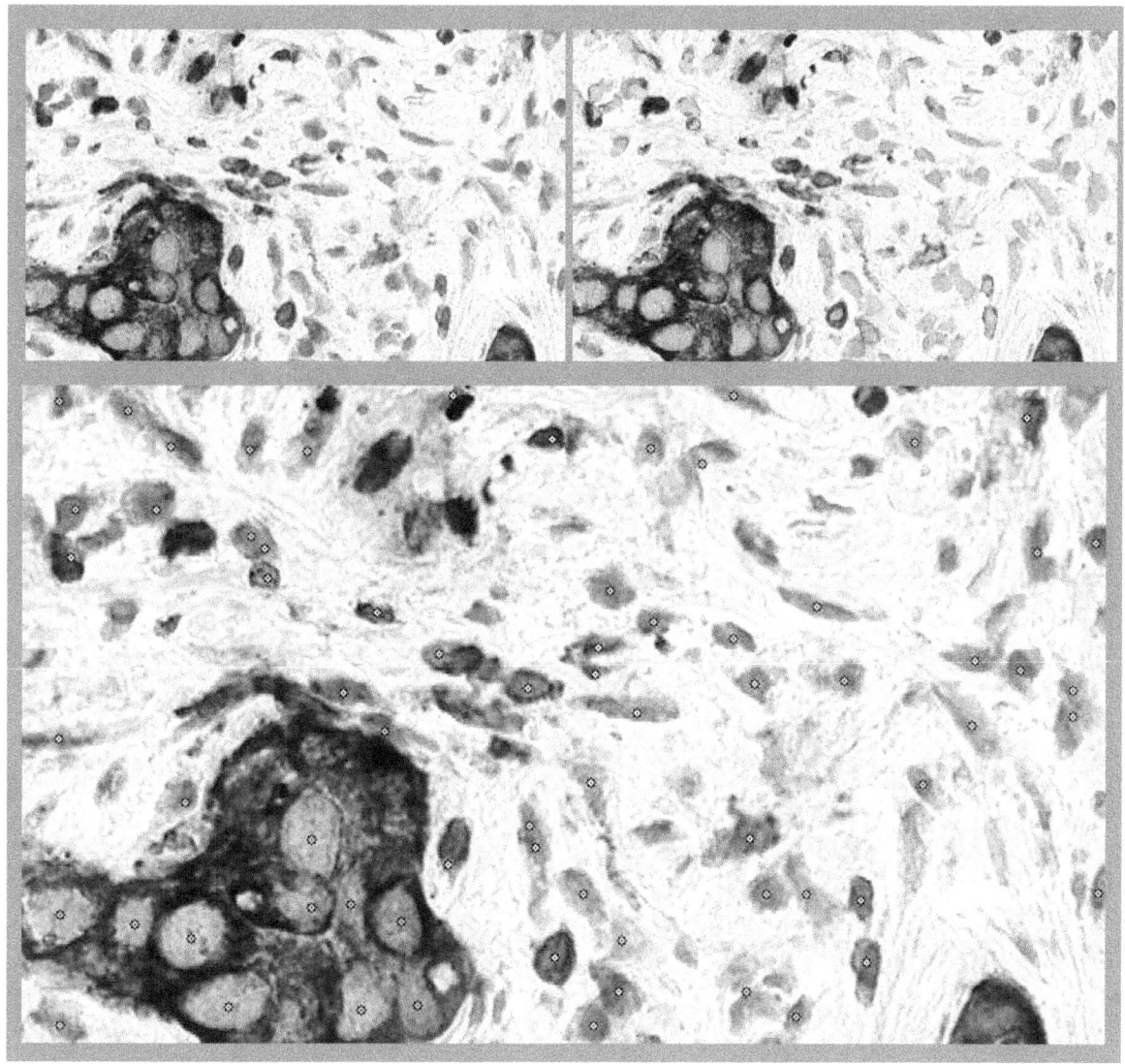

GemIdent identifying cancer cells, T-cells, and background nuclei in a microscopic image

16.7   External links

- GemIdent's homepage

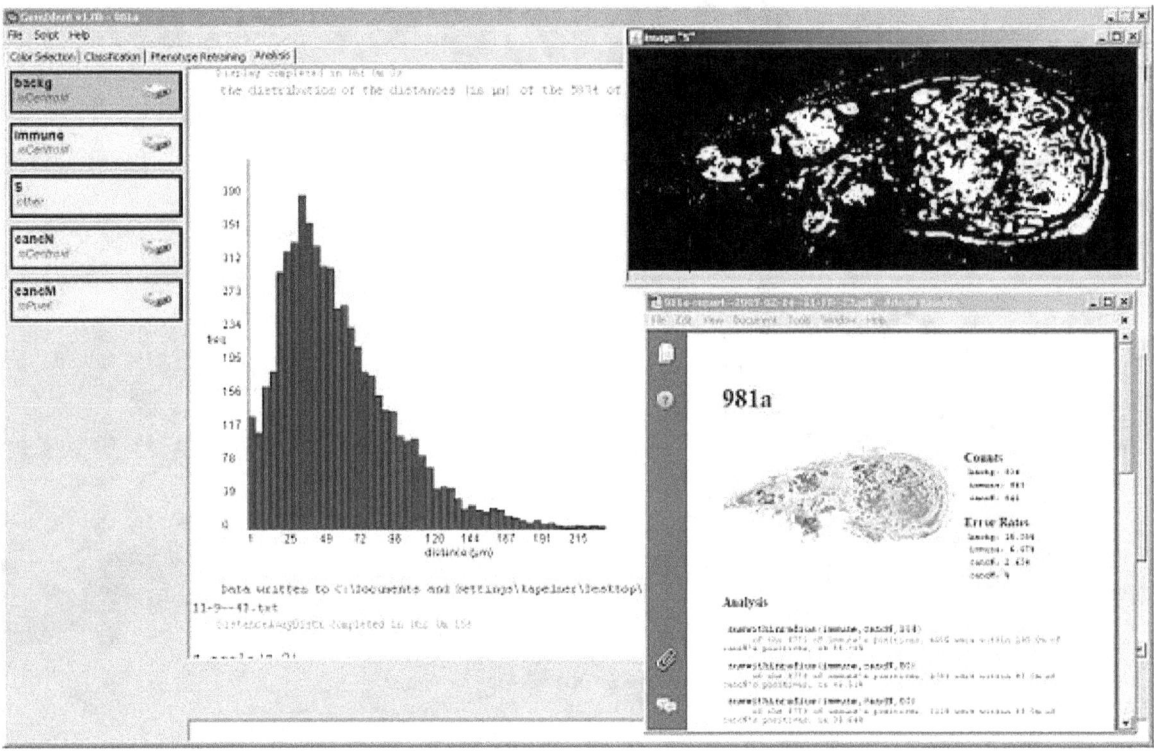

GemIdent analyzing results using data analysis and visualization tools

Chapter 17

GIMIAS

GIMIAS is a workflow-oriented environment focused on biomedical image computing and simulation. The open source framework is extensible through plug-ins and is focused on building research and clinical software prototypes. Gimias has been used to develop clinical prototypes in the fields of cardiac imaging and simulation, angiography imaging and simulation, and neurology[1][2]

GIMIAS is being funded by several national and international projects like cvREMOD, euHeart or VPH NoE.[3]

17.1 About GIMIAS

GIMIAS stands for Graphical Interface for Medical Image Analysis and Simulation. GIMIAS provides a graphical user interface with all main data IO, visualization and interaction functions for images, meshes and signals. GIMIAS features include:

- DICOM browser and PACS connection

- Support for different imaging modalities

- Biomedical data visualization in 2D and 3D: multiplanar reformation, ortho slice view, multi slice view, volume rendering, X-ray rendering, maximum intensity projection

- Several input and output formats: DICOM, vtk, stl, Nifty, Analyze.

- Movie control: play, pause, speed control

- Multiple data objects: 2D DICOM images, 3D images, surface meshes, volumetric meshes, signals or annotations

- Image and surface mesh annotations: landmarks, measurements and regions of interest

- Clinical workflow navigation that can help the user to navigate from patient data to useful information for patient treatment.[4]

- Other additional tools for image segmentation, mesh manipulation and signal navigation.[5]

GIMIAS is a development framework that allows developers to create their own medical applications using different plug-ins that can be dynamically loaded and combined. The prototypes developed on GIMIAS can be verified by end users in real scenarios and with real data at early development stages.[6]

Is developed using C++ language, has a plug-in architecture, and is cross-platform by means of the standard CMake tool. Is possible to integrate new libraries using CSnake tool and is based on common open source libraries like VTK,

ITK, MITK, BOOST and wxWidgets. A plug-in can extend the framework adding new processing components, GUI components like toolbars or windows, new data processing types or new rendering libraries.[7]

GIMIAS supports several types of plug-ins, starting from a simple DLL, a 3D Slicer compatible command line plug-in or a more complex GIMIAS plug-in with customized graphical interface. Automated GUI generation and extensible data object model allow to share plug-ins with other frameworks and empower interoperability.

The software is available on Windows and Linux, 64-bit and 32-bit.[8]

17.2   Image gallery

17.3   History

Initial versions of the open source framework was released by the end of 2009 (GIMIAS 0.6.15 was released on October 2009).[9]

In 2010, more effort was done to empower the open source framework itself, providing more functionality like workflow manager, 3D Slicer plug-in compatibility, signal viewer and customizable views. GIMIAS version 0.8.1, 1.0.0, 1.1.0 and 1.2.0 were released during this year.

GIMIAS Team have collaborated with:

- cmgui team: to trial the use of the interim cmgui API from the GIMIAS software platform[10][11]

- CTK group[12]

- B3C group (MAF)[13]

GIMIAS is one of the tools used in the Virtual Physiological Human.[14]

17.4   Clinical Prototypes

- **AngioLab** is a software tool developed within the GIMIAS framework and is part of a more ambitious pipeline for the integrated management of cerebral aneurysms. AngioLab currently includes four plug-ins: angio segmentation, angio morphology virtual stenting and virtual angiography. In December 2009, 23 clinicians completed an evaluation questionnaire about AngioLab. This activity was part of a teaching course held during the 2nd European Society for Minimally Invasive Neurovascular Treatment (ESMINT) Teaching Course held at the Universitat Pompeu Fabra, Barcelona, Spain. The Automated Morphological Analysis (angio morphology plug-in) and the Endovascular Treatment Planning (stenting plug-in) were evaluated. In general, the results provided by these tools were considered as relevant and as an emerging need in their clinical field.[15][16]

- **CardioLab**: The CardioLab suite for GIMIAS allows to perform an entire workflow from medical images to characterization and quantification of myocardial diseases and Cardiac Resynchronization Therapy (CRT) planning.[17]

- **FocusDET**: Accurate localization of epileptogenic foci in intractable partial epilepsy is essential for assessing the possibility of surgery as a treatment. A specific software package was developed to locate the epileptogenic focus using Ictal and Inter-ictal SPECT images and MRI employing the SISCOM methodology. FocusDET was developed using GIMIAS facilities.[18]

- **QuantiDopa** is a software that allows to perform a semiautomatic quantification of the striatal uptake in neurotransmission SPECT studies of the dopaminergic system.[19]

17.5   References

[1] I. Larrabide, P. Omedas, Y. Martelli, X. Planes, M. Nieber, J. A. Moya, C. Butakoff, R. Sebastián, O. Camara, M. De Craene, B. Bijnens, A.F. Frangi, GIMIAS: An open source framework for efficient development of research tools and clinical prototypes, in Functional Imaging and Modeling of the Heart, 417-426, 2009.

[2] "VPH Requirements and Technology Assessment Exercise". Virtual Physiological Human Network of Excellence. p. 95.

[3] "cvREMOD web site".

[4] ahc. "I do imaging".

[5] P. Omedas,Y. Martelli, I. Larrabide, B. H. Bijnens, A. F. Frangi. "Advance Tool for Visualization of Multi-modal and Multi-scale Cardiac Data" (PDF). UPF. p. 42.

[6] "GIMIAS Home Page".

[7] GIMIAS Team. "GIMIAS architecture". slideshare.

[8] Hamza Emadeen Mousa. "GIMIAS Medical Image Analysis and Simulation Solution for Windows and Linux". goomedic.

[9] GIMIAS Team. "GIMIAS on SourceForge". SourceForge.

[10] "CMGUI and Data Fusion". Virtual Physiological Human Network of Excellence. p. 30.

[11] "Introduction to cmgui". Auckland Bioengineering Institute.

[12] "CTK-Hackfest-May-2010".

[13] "B3C collaboration" (PDF).

[14] "Virtual Physiological Human".

[15] M.C. Villa-Uriol, I. Larrabide, J.M. Pozo, H. Bogunovic, P. Omedas, V. Barbarito, L. Carotenuto, C. Riccobene, X. Planes, Y. Martelli, A.J. Geers and A.F. Frangi, AngioLab: Integrated technology for patient-specific management of intracranial aneurysms, 32nd Annual International Conference of the IEEE Engineering in Medicine and Biology Society (EMBS), Buenos Aires, Argentina, 2010

[16] "GIMIAS toolchain for aneurysm rupture". vph-noe.

[17] "Building a pipeline for in-silico modelling of cardiac resynchronization therapy". vph-noe. p. 11.

[18] B. Martí, Ó. Esteban, X. Planes, P. Omedas, G. Wollny, A. Cot, X. Setoain, A. Frangi, M. Ledesma-Carbayo, J. Pavia (2009). "FocusDET: A software tool to locate epileptogenic foci in intractable partial epilepsy". *ENAM*.

[19] "VPH2010 Conference Presentation Schedule". vph-noe.

17.6   External links

- GIMIAS Home Page

- MITK

Chapter 18

Ginkgo CADx

Ginkgo CADx is a multiplatform (Windows, Linux,[1] Mac OS X) DICOM viewer (*.dcm) and dicomizer (convert different files to DICOM). Ginkgo CADx is licensed under LGPL license, being an open source project with an Open core approach. The goal of Ginkgo CADx project is to develop an open source professional DICOM workstation.[2]

Ginkgo CADx also has a professional version with a private license model called Ginkgo CADx Pro.

18.1   About Ginkgo CADx

The main features of this software are:

- Fully functional DICOM viewer and DICOM workstation with PACS support (C-FIND, C-GET, C-MOVE, C-STORE...).

- Dicomizator: Ginkgo CADx can convert from JPEG, PNG, GIF, TIFF images and PDF documents to DICOM files.

- Cross-platform (Windows, Linux, Mac OS X).

- Standard DICOM tools: rulers, angle, window level, cinema mode...

- Support a large number of different DICOM modalities as CT, MR, ECG, PET, XC, SC, XA, MG...

- Intuitive graphical interface customizable by user profiles.

- Easily integrable with third party systems using HL7 messages, DICOM and IHE compliant workflows.

Ginkgo CADx is developed using C++ language, has a plug-in architecture, and is cross-platform by means of the standard CMake tool. Is based on common open source libraries like VTK,[3] ITK, and wxWidgets.

18.2   Image gallery

18.3   History

Ginkgo CADx project started in 2009[4] with the aim to create an interactive, universal, homogeneous, open-source and cross-platform CADx environment. Developed by MetaEmotion, the first public version (2.0.2.0) of Ginkgo CADx was released in 2010.

Ginkgo CADx team have collaborated with:

- Sacyl Public healthcare system of Castilla y León.

- GNUmed team. Packaging and compiling Ginkgo CADx for Ubuntu, Debian, Fedora and SUSE.

18.4 See also

18.5 References

[1] Ginkgo CADx package in Ubuntu

[2] The technology behind medical imagery. The Hindu newspaper.

[3] Ginkgo CADx: Open Source DICOM CADx Environment. Kitware Source. April 2012.

[4] I Do Imaging: Ginkgo CADx outline

18.6 External links

- Ginkgo CADx webpage

- Ginkgo CADx Forums

Chapter 19

Google Goggles

Not to be confused with Google Glass.

Google Goggles is an image recognition mobile app developed by Google.[1] It is used for searches based on pictures taken by handheld devices. For example, taking a picture of a famous landmark searches for information about it, or taking a picture of a product's barcode searches for information on the product.[2]

19.1 History

Google Goggles was developed for use on Google's Android operating systems for mobile devices. While initially only available in a beta version for Android phones, Google announced its plans to enable the software to run on other platforms, notably iPhone and BlackBerry devices.[3] Google did not discuss a non-handheld format. On 5 October 2010, Google announced availability of Google Goggles for iPhone and iPad devices running iOS 4.0.[4]

In a May 2014 update to Google Mobile for iOS, the Google Goggles feature was removed due to being "of no clear use to too many people."[5]

19.2 Uses

The system could identify various labels or landmarks, allowing users to learn about such items without needing a text-based search. The system could identify products barcodes or labels that allow users to search for similar products and prices, and save codes for future reference, similar to the failed CueCat of the late '90s, but with more functionality.[2] The system also recognized printed text and use optical character recognition (OCR) to produce a text snippet, and in some cases even translate the snippet into another language.[2]

19.2.1 Metropolitan Museum of Art

The Metropolitan Museum of Art announced in December 2011 its collaboration with Google to use Google Goggles for providing information about the artworks in the museum through direct links to the website of the Metropolitan Museum of Art.[6]

19.3 Current version

The final version of Google Goggles was 1.9[7] which adds several new features and improves both quality and ease of use. Goggles is specifically developed to run on mobile devices running the Android operating system and can be installed using Google Play (formerly Android Market).[2]

Although developed for Android there was an iPhone version, as part of the Google Search app, available from the iTunes Store or App Store. Goggles for iPhone required iPhone 3GS or iPhone 4 or[8] iOS 4.0 or higher to run.[2]

In January 2011, version 1.3 was released; it could solve Sudoku puzzles.[9]

In late August 2012, Google launched an update to its Google Goggles app, version 1.9. This update put an emphasis on helping users shop by including improved product recognition and new recommendations that help users browse similar products.[10]

Earlier versions of the Android app were able to load pictures from the phone's gallery, which had been removed in version 1.9.2; however, it could be worked around by sharing the image to the Goggles app from a file browser.

19.4 Platform

Google product manager Shailesh Nalawadi indicated that Google wanted Goggles to be an application platform, much like Google Maps, not just a single product.

19.5 See also

- Google Glass
- Google Cardboard, a virtual reality headset developed by Google
- Google Mobile
- Nokia Point & Find
- Bing Vision

19.6 References

[1] Google Mobile

[2] Google mobile: Mainpage on Google Goggles, visited 8 December 2010

[3] PCWorld: Goggles will reach other platforms

[4] "Open your eyes: Google Goggles now available on iPhone in Google Mobile App". *Google Mobile Blog.* October 5, 2010.

[5] https://productforums.google.com/d/msg/websearch/ZIiPQRDTSEQ/Yh-sXzN36ZAJ

[6] *Metropolitan Museum Enhances Online Access to Its Collections with Google Goggles.* New York, December 16, 2011; Thomas P. Campbell: Google Goggles (New York, December 16, 2011): *I'm pleased to announce a new collaboration with Google that lets you take a picture of a work of art with your mobile device and link straight to more information on metmuseum.org.*

[7] Google Goggles Release Notes, visited 13 June 2011

[8] "Google Help". Support.google.com. Retrieved 2013-06-16.

[9] T3 website: Goggles can now solve sudoku puzzles, 11 January 2011. Visited 6 August 2011

[10] Weber, Harrison. August 23, 2012. "Google Goggles' latest update makes it easier to shop IRL"

19.7 External links

- Official website

Chapter 20

Ilastik

ilastik[1] is a user-friendly free open source software for image classification and segmentation. No previous experience in image processing is required to run the software.

20.1 Features

ilastik allows user to annotate an arbitrary number of classes in images with a mouse interface. Using these user annotations and the generic (nonlinear) image features, the user can train a random forest classifier. ilastik has a CellProfiler module to use ilastik classifiers to process images within a CellProfiler framework.

20.2 History

ilastik was first released in 2011 by scientists at the Heidelberg Collaboratory for Image Processing (HCI), University of Heidelberg.

20.3 Application

- The Interactive Learning and Segmentation Toolkit
- Carving[2][3]
- Cell classification and neuron classification[4]
- Synapse detection

20.4 Resources

ilastik project is hosted on GitHub. It is a collaborative project, any contributions such as comments, bug reports, bug fixes or code contributions are welcome.

20.5 References

[1] Sommer, C; Straehle C; Koethe U; Hamprecht FA (2011). "ilastik: Interactive Learning and Segmentation Toolkit". *IEEE International Symposium on Biomedical Imaging*: 230–33. doi:10.1109/ISBI.2011.5872394.

[2] Straehl, C; Köthe U; Briggman K; Denk W; Hamprecht FA (2012). "Seeded watershed cut uncertainty estimators for guided interactive segmentation". *CVPR*.

[3] Straehle, CN; Köthe U; Knott G; Hamprecht FA (2011). "Carving: scalable interactive segmentation of neural volume electron microscopy images". *MICCAI* **14** (Pt 1): 653–60. doi:10.1007/978-3-642-23623-5_82. PMID 22003674.

[4] Kreshuk, A; Straehle CN; Sommer C; Koethe U; Cantoni M; et al. (2011). "Automated detection and segmentation of synaptic contacts in nearly isotropic serial electron microscopy images". *Automated Detection and Segmentation of Synaptic Contacts in Nearly Isotropic Serial Electron Microscopy Images* **6** (10): e24899. doi:10.1371/journal.pone.0024899. PMC 3198725. PMID 22031814.

20.6   External links

- Official website

- ilastik on GitHub

Chapter 21

ILNumerics.Net

ILNumerics is a mathematical class library for Common Language Infrastructure (CLI) developers. It simplifies the implementation of an array of numerical algorithms. ILNumerics was designed to help developers create distribution-ready applications. Interfaces of existing algebra systems were often found to be less effective, when it comes to distribution/integration into existing projects; therefore, ILNumerics does not come with an interpreter but directly utilizes features of modern development environments and programming languages like C#.

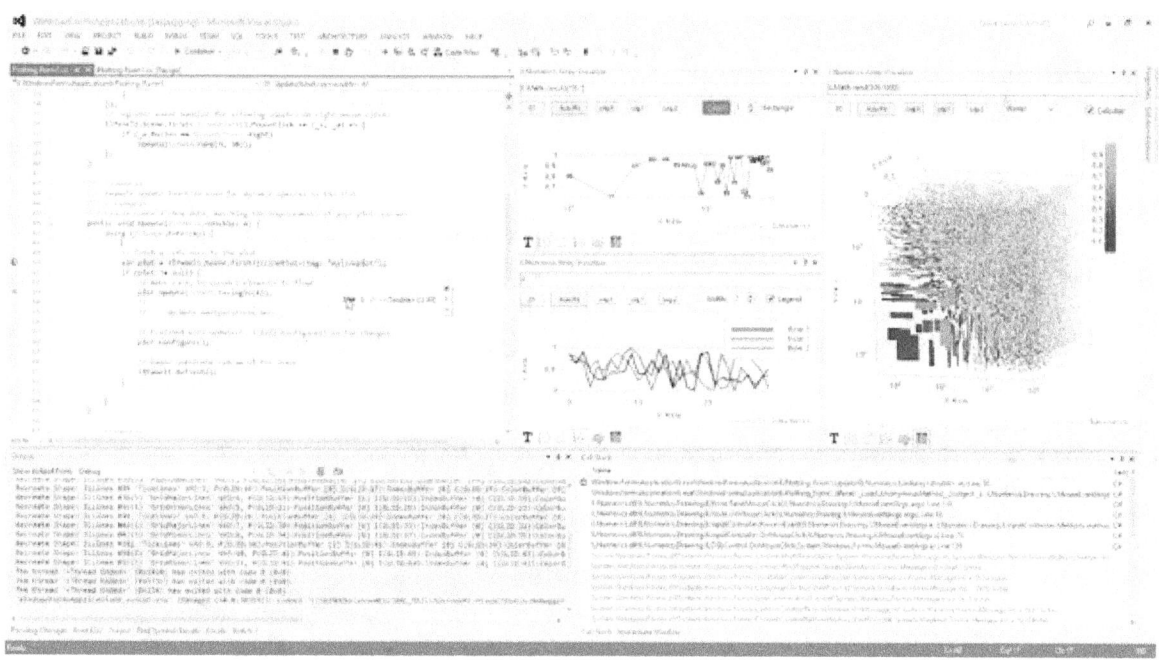

Shows the use of ILNumerics Array Visualizer while debugging in a multilanguage code project in Visual Studio.

21.1 Features

N-dimensional arrays, complex numbers, linear algebra, FFT and plotting controls (2D and 3D) help developing algorithms on every platform the CLI runs on. Developers formulate computational algorithms directly in their favorite CLI language - avoiding the need for interfacing 3rd party mathematical frameworks. The syntax is vastly compatible to well known and established mathematical programs like MATLAB and GNU Octave. Due to its strong type safety algorithms developed that way are more stable and robust at run time. The library is the only math library so far, which takes the

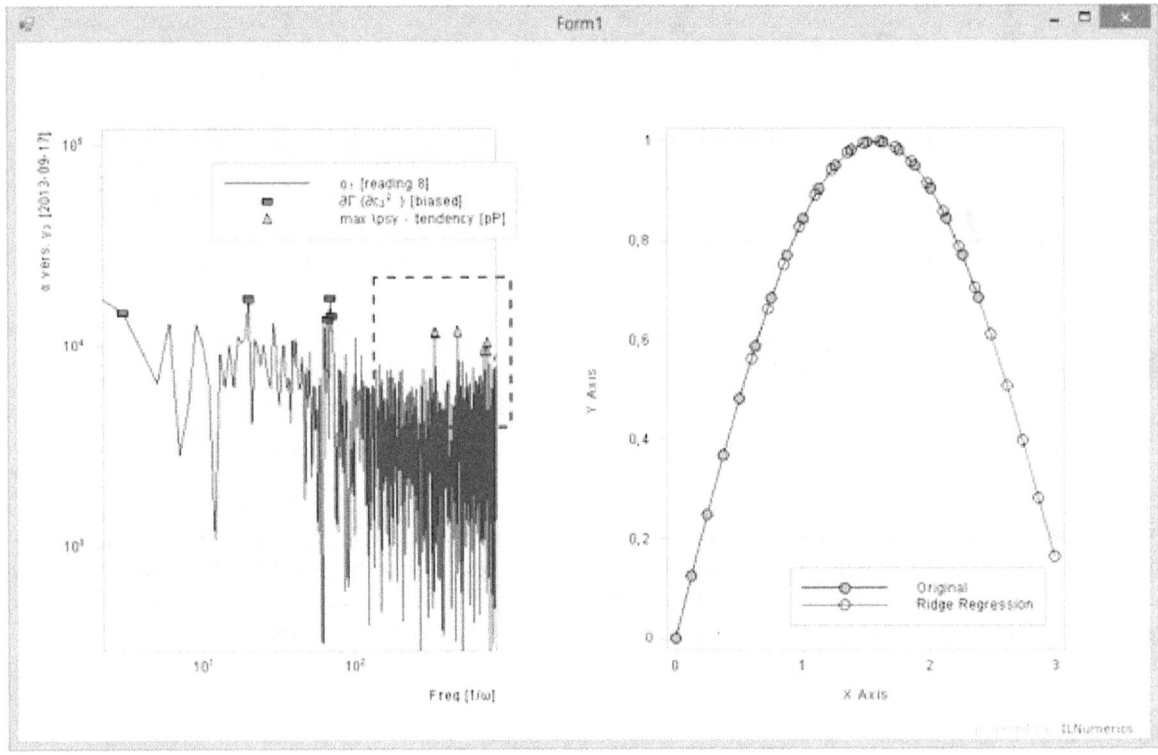

Shows the use of ILNumerics ILPanel for 2D plot creation directly inside Windows Forms applications.

characteristics of the CLI into account and therefore achievers better execution performance than its competitors.

Based on the foundation of efficient n-dimensional arrays, an optimization toolbox, high level HDF5 APIs and a number of high level statistics and machine learning algorithms are provided. ILNumerics allows the rapid development of interactive, production ready 2D and 3D dynamic visualizations, based on scene graphs and rendered on Windows Forms, OpenGL, GDI+ and SVG.

ILNumerics features several convenient debug options. The ILNumerics Array Visualizer is integrated into Visual Studio and allows the graphical inspection of mathematical objects while stepping through the code. Due to its developer efficiency, ILNumerics is known as RAD tool for technical application development.

21.2 Performance

Since ILNumerics comes as a CLI assembly, it targets Common Language Infrastructure (CLI) applications. Just like Java - those frameworks are often criticized for not being suitable for numerical computations. Reasons are the memory management by a garbage collector and the intermediate language execution. Nevertheless, due to efficient memory management (pooling), the performance of ILNumerics algorithms beat the speed of many competing frameworks by factors.[1] Linear algebra routines rely on processor specific optimized versions of LAPACK and BLAS, which further increases performance and reliability of computational results. All internal functions are parallelized. The efficiency allows the use for 'numbercrunching' applications, which would otherwise only be suitable for Fortran - yet providing much higher implementational convenience.

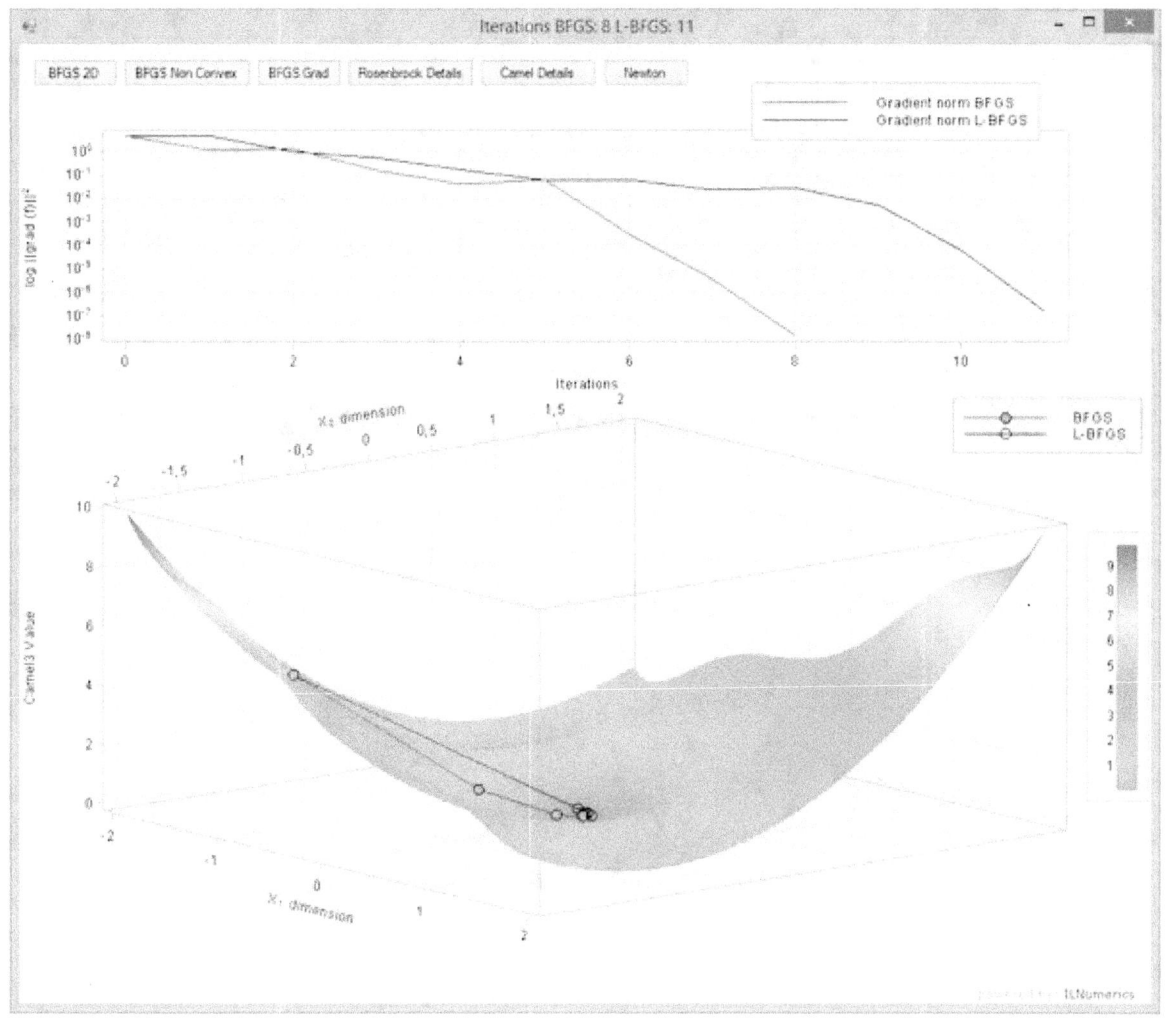

Shows the use of ILNumerics Optimization Toolbox for minimization in a Windows Forms application.

21.3   History

ILNumerics started in 2006 and serves its community with high performance fundamental math classes since. In 2007 ILNumerics won the BASTA! Innovation Awards 2007[2] as most innovative .NET project in Germany, Switzerland and Austria. After 6 years of open source development, the project added a closed source, proprietary license in 2011, aiming business and academic developers at the same time.

21.4   See also

- List of numerical libraries

- Comparison of numerical analysis software

21.5   References

[1] http://ilnumerics.net/blog/fast-faster-performance-comparison-c-ilnumerics-fortran-matlab-and-numpy-part-i/

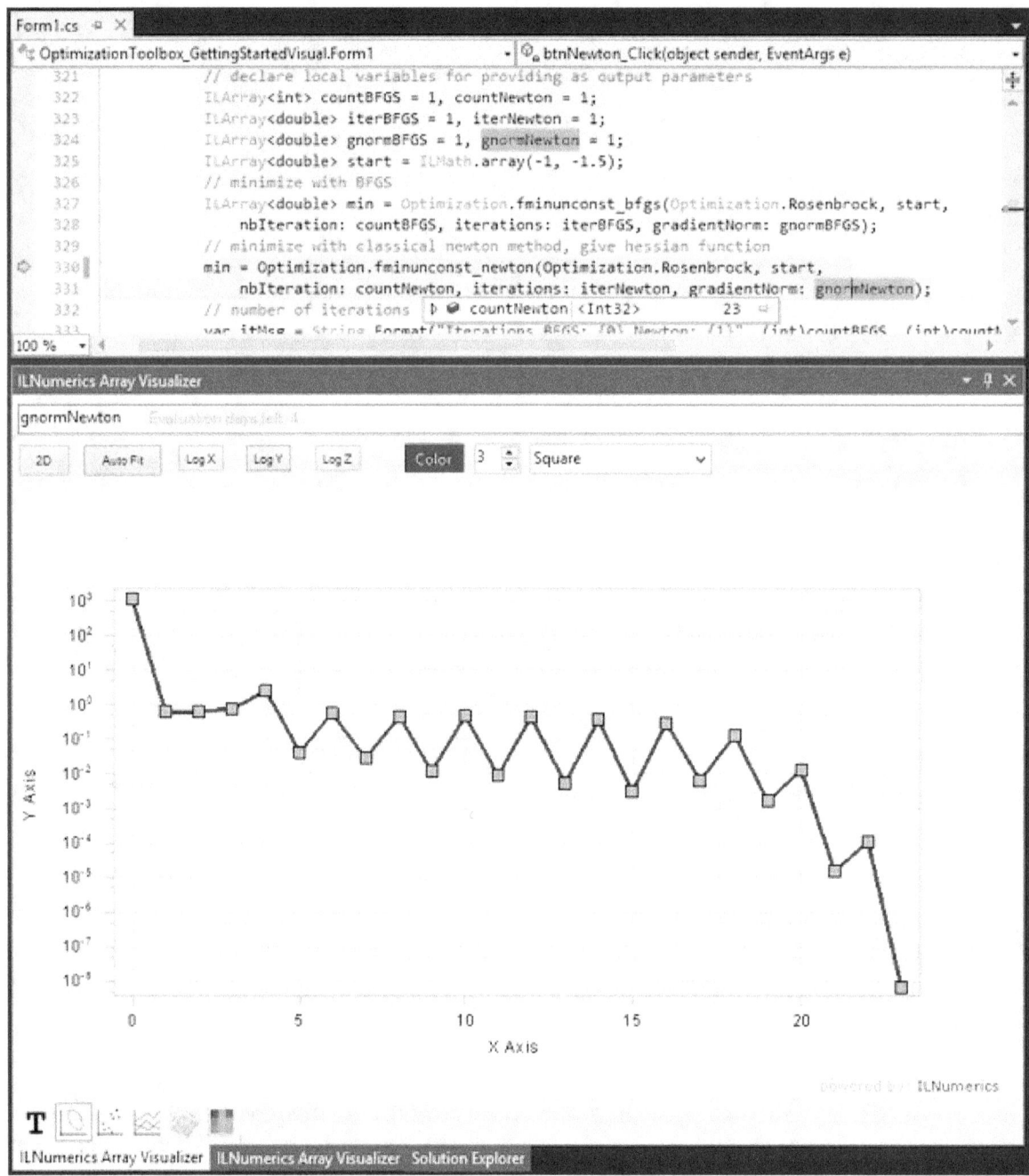

Shows the use of ILNumerics Array Visualizer while debugging in Visual Studio.

[2] BASTA! Innovation Award 2007

21.6 External links

- ILNumerics website (proprietary software)

Chapter 22

ImageNets

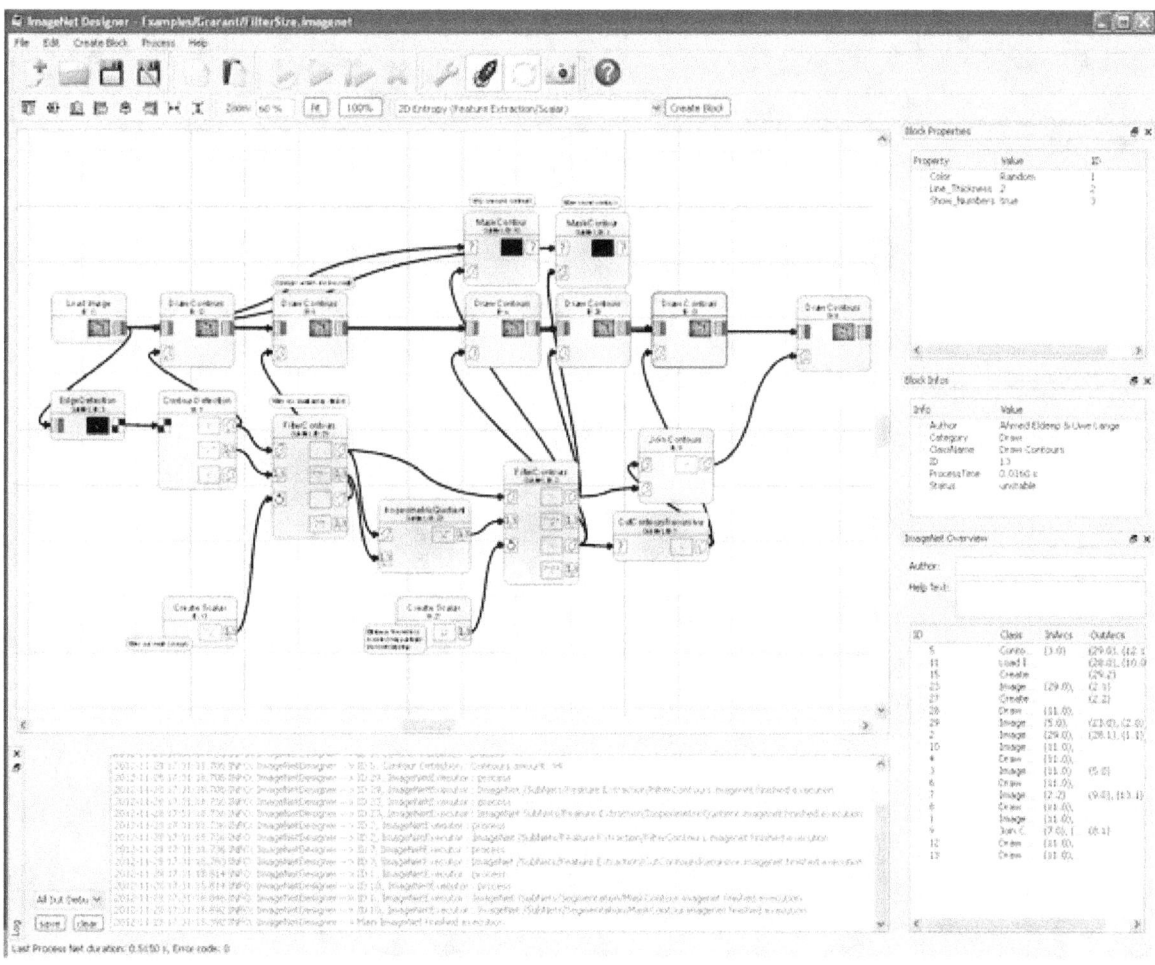

ImageNet Designer

ImageNets is an open source and platform independent (Windows & Linux) framework for rapid prototyping of Machine Vision algorithms. With the GUI *ImageNet Designer*, no programming knowledge is required to perform operations on images. A configured ImageNet can be loaded and executed from C++ code without the need for loading the *ImageNet Designer* GUI to achieve higher execution performance. Originally, ImageNets was developed for the Care-Providing Robot FRIEND [1][2] but it can be used for a wide range of computer vision applications.

72

ImageNets was developed by the Institute of Automation, University of Bremen, Germany. The software was first publicly released in 2010.

22.1 References

[1] Prenzel, O., Lange, U., Kampe, H., Martens, C., Gräser, A. (2012). *Programming of Intelligent Service Robots with the Process Model "FRIEND::Process" and Configurable Task-Knowledge.* Croatia: Intech. pp. 529–552. ISBN 978-953-307-941-7.

[2] Lange, U., Kampe, H. Gräser, A. (2012). *Fast Development of a Robust Calibration Method for Service Robot FRIEND using ImageNets* (PDF). Germany: Shaker. pp. 25–36. ISBN 978-3-8440-1322-1.

22.2 External links

- ImageNets homepage

- Download ImageNets here

Chapter 23

Insight Segmentation and Registration Toolkit

ITK is a cross-platform, open-source application development framework widely used for the development of image segmentation and image registration programs. Segmentation is the process of identifying and classifying data found in a digitally sampled representation. Typically the sampled representation is an image acquired from such medical instrumentation as CT or MRI scanners. Registration is the task of aligning or developing correspondences between data. For example, in the medical environment, a CT scan may be aligned with an MRI scan in order to combine the information contained in both.

ITK was developed with funding from the National Library of Medicine (U.S.) as an open resource of algorithms for analyzing the images of the Visible Human Project. ITK stands for **The Insight Segmentation and Registration Toolkit**. The toolkit provides leading-edge segmentation and registration algorithms in two, three, and more dimensions. ITK uses the CMake build environment to manage the configuration process. The software is implemented in C++ and it is wrapped for Python and Java. This enables developers to create software using a variety of programming languages. ITK's C++ implementation style is referred to as generic programming (i.e., using templated code). Such C++ templating means that the code is highly efficient, and that many software problems are discovered at compile-time, rather than at run-time during program execution.

23.1 Introduction

23.1.1 Origins

In 1999 the US National Library of Medicine of the National Institutes of Health awarded a three-year contract to develop an open-source registration and segmentation toolkit, which eventually came to be known as the Insight Toolkit (ITK). ITK's NLM Project Manager was Dr. Terry Yoo, who coordinated the six prime contractors who made up the Insight Software Consortium. These consortium members included the three commercial partners GE Corporate R&D, Kitware, Inc., and MathSoft (the company name is now Insightful); and the three academic partners University of North Carolina (UNC), University of Tennessee (UT), and University of Pennsylvania (UPenn). The Principal Investigators for these partners were, respectively, Bill Lorensen at GE CRD, Will Schroeder at Kitware, Vikram Chalana at Insightful, Stephen Aylward with Luis Ibáñez at UNC (both of whom subsequently moved to Kitware), Ross Whitaker with Josh Cates at UT (both now at Utah), and Dimitris Metaxas at UPenn (Dimitris Metaxas is now at Rutgers University). In addition, several subcontractors rounded out the consortium including Peter Ratiu at Brigham & Women's Hospital, Celina Imielinska and Pat Molholt at Columbia University, Jim Gee at UPenn's Grasp Lab, and George Stetten at University of Pittsburgh.

23.1.2 Technical details

ITK is an open-source software toolkit for performing registration and segmentation. Segmentation is the process of identifying and classifying data found in a digitally sampled representation. Typically the sampled representation is an image acquired from such medical instrumentation as CT or MRI scanners. Registration is the task of aligning or developing correspondences between data. For example, in the medical environment, a CT scan may be aligned with an MRI scan in order to combine the information contained in both.

ITK is implemented in C++. ITK is cross-platform, using the CMake build environment to manage the compilation process. In addition, an automated wrapping process generates interfaces between C++ and interpreted programming languages such as Java and Python. This enables developers to create software using a variety of programming languages. ITK's implementation employs the technique of generic programming through the use of C++ templates.

Because ITK is an open-source project, developers from around the world can use, debug, maintain, and extend the software. ITK uses a model of software development referred to as extreme programming. Extreme programming collapses the usual software creation methodology into a simultaneous and iterative process of design-implement-test-release. The key features of extreme programming are communication and testing. Communication among the members of the ITK community is what helps manage the rapid evolution of the software. Testing is what keeps the software stable. In ITK, an extensive testing process (using CDash) is in place that measures the quality on a daily basis. The ITK Testing Dashboard is posted continuously, reflecting the quality of the software.

23.1.3 Developers and Contributors

The Insight Toolkit was initially developed by six principal organizations

- Kitware

- GE Corporate R&D

- Insightful

- University of North Carolina at Chapel Hill

- University of Utah

- University of Pennsylvania

and three subcontractors

- Harvard Brigham & Women's Hospital

- University of Pittsburgh

- Columbia University

After its inception the software continued growing with contributions from other institutions including

- University of Iowa

- Georgetown University

- Stanford University

- King's College of London

- Creatis INSA

23.1.4 Funding

The funding for the project is from the National Library of Medicine at the National Institutes of Health. NLM in turn was supported by member institutions of NIH (see sponsors).

The goals for the project include the following:

- Support the Visible Human Project.

- Establish a foundation for future research.

- Create a repository of fundamental algorithms.

- Develop a platform for advanced product development.

- Support commercial application of the technology.

- Create conventions for future work.

- Grow a self-sustaining community of software users and developers.

The source code of the Insight Toolkit is distributed under an Apache 2.0 License (as approved by the Open Source Initiative)

The philosophy of Open Source of the Insight Toolkit was extended to support Open Science, in particular by providing Open Access to publications in the domain of Medical Image Processing. These publications are made freely available through the Insight Journal

23.1.5 Community Participation

Because ITK is an open-source system, anybody can make contributions to the project. A person interested in contributing to ITK can take the following actions

1. Read the ITK Software Guide. (This book can be purchased from Kitware's store.)

2. Read the instructions on how to contribute classes and algorithms to the Toolkit via submissions to the Insight Journal

3. Obtain access to ITK's Gerrit Code Review instance.

4. Follow the Git contribution instructions.

5. Join the insight-users list. Subscriptions to the list are open to everybody.

Anyone can submit a patch, and write access to the repository is not necessary to get a patch merged or retain authorship credit. For more information, see the ITK Bar Camp documentation on how to submit a patch.

23.1.6 Copyright and License

ITK is copyrighted by the Insight Software Consortium, a non-profit alliance of organizations and individuals interested in supporting ITK. Starting with ITK version 3.6, the software is distributed under a BSD open-source license. It allows use for any purpose, with the possible exception of code found in the patented directory, and with proper recognition. The full terms of the copyright and the license are available at www.itk.org/ITK/project/license.html. Version 4.0 uses Apache 2.0 License.

The licensed was changed to Apache 2.0 with version 4.0 to adopt a modern license with patent protection provisions. From version 3.6 to 3.20, a simplified BSD license was used. Versions of ITK previous to ITK 3.6 were distributed under a modified BSD License. The main motivation for adopting a BSD license starting with ITK 3.6, was to have an OSI-approved license.

23.2 Technical Summary

The following sections summarize the technical features of the NLM's Insight ITK toolkit. Design Philosophy The following are key features of the toolkit design philosophy.

- The toolkit provides data representation and algorithms for performing segmentation and registration. The focus is on medical applications; although the toolkit is capable of processing other data types.

- The toolkit provides data representations in general form for images (arbitrary dimension) and (unstructured) meshes.

- The toolkit does not address visualization or graphical user interface. These are left to other toolkits (such as VTK, VisPack, 3DViewnix, MetaImage, etc.)

- The toolkit provides minimal tools for file interface. Again, this is left to other toolkits/libraries to provide.

- Multi-threaded (shared memory) parallel processing is supported.

- The development of the toolkit is based on principles of extreme programming. That is, design, implementation, and testing is performed in a rapid, iterative process. Testing forms the core of this process. In Insight, testing is performed continuously as files are checked in, and every night across multiple platforms and compilers. The ITK testing dashboard, where testing results are posted, is central to this process.

23.2.1 Architecture

The following are key features of the toolkit architecture.

- The toolkit is organized around a data-flow architecture. That is, data is represented using data objects which are in turn processed by process objects (filters). Data objects and process objects are connected together into pipelines. Pipelines are capable of processing the data in pieces according to a user-specified memory limit set on the pipeline.

- Object factories are used to instantiate objects. Factories allow run-time extension of the system.

- A command/observer design pattern is used for event processing.

23.2.2 Implementation Philosophy

The following are key features of the toolkit implementation philosophy.

- The toolkit is implemented using generic programming principles. Such heavily templated C++ code challenges many compilers; hence development was carried out with the latest versions of the MSVC, Sun, gcc, Intel, and SGI compilers.

- The toolkit is cross-platform (Unix, Windows and Mac OS X).

- The toolkit supports multiple language bindings, including such languages as Tcl, Python, and Java. These bindings are generated automatically using an auto-wrap process.

- The memory model depends on "smart pointers" that maintain a reference count to objects. Smart pointers can be allocated on the stack, and when scope is exited, the smart pointers disappear and decrement their reference count to the object that they refer to.

23.2.3   Build Environment

ITK uses the CMake (cross-platform make) build environment. CMake is an operating system and compiler independent build process that produces native build files appropriate to the OS and compiler that it is run with. On Unix CMake produces makefiles and on Windows CMake generates projects and workspaces.

23.2.4   Testing Environment

ITK supports an extensive testing environment. The code is tested daily (and even continuously) on many hardware/operating system/compiler combinations and the results are posted daily on the ITK testing dashboard. We use Dart to manage the testing process, and to post the results to the dashboard.

23.2.5   Background References: C++ Patterns and Generics

ITK uses many advanced design patterns and generic programming. You may find these references useful in understanding the design and syntax of Insight.

- Design Patterns. by Erich Gamma, Richard Helm, Ralph Johnson, John Vlissides, Grady Booch

- Generic Programming and the Stl : Using and Extending the C++ Standard Template Library (Addison-Wesley Professional Computing Series) by Matthew H. Austern

- Advanced C++ Programming Styles and Idioms by James O. Coplien

- C/C++ Users Journal

- C++ Report

23.3   Examples

23.3.1   Gaussian-smoothed Image Gradient

```
#include "itkImage.h" int main() { typedef itk::Image< unsigned char, 3 > ImageType; typedef itk::ImageFileReader< Im-
ageType > ReaderType; typedef itk::ImageFileWriter< ImageType > WriterType; typedef itk::GradientRecursiveGaussianImageFilter<
ImageType, ImageType > FilterType; ReaderType::Pointer reader = ReaderType::New(); WriterType::Pointer writer =
WriterType::New(); reader->SetFileName("lungCT.dcm"); writer->SetFileName("smoothedLung.hdr"); FilterType::Pointer
filter = FilterType::New(); filter->SetInput( reader->GetOutput() ); writer->SetInput( filter->GetOutput() ); filter->SetSigma();
try { writer->Update(); } catch( itk::ExceptionObject & excp ) { std::cerr << excp << std::endl; return EXIT_FAILURE;
} }
```

23.3.2   Region Growing Segmentation

```
#include "itkImage.h" int main() { typedef itk::Image< signed short, 3 > InputImageType; typedef itk::Image< unsigned
char, 3 > OutputImageType; typedef itk::ImageFileReader< InputImageType > ReaderType; typedef itk::ImageFileWriter<
OutputImageType > WriterType; typedef itk::ConnectedThresholdImageFilter< InputImageType, OutputImageType >
FilterType; ReaderType::Pointer reader = ReaderType::New(); WriterType::Pointer writer = WriterType::New(); reader-
>SetFileName("brain.dcm"); writer->SetFileName("whiteMatter.hdr"); FilterType::Pointer filter = FilterType::New();
filter->SetInput( reader->GetOutput() ); writer->SetInput( filter->GetOutput() ); filter->SetMultiplier(2.5); ImageType::IndexType
seed; seed[0] = 142; seed[1] = 97; seed[2] = 63; filter->AddSeed( seed ); try { writer->Update(); } catch( itk::ExceptionObject
& excp ) { std::cerr << excp << std::endl; return EXIT_FAILURE; } }
```

23.4 Additional information

23.4.1 Resources

A number of resources are available to learn more about ITK.

- The ITK web pages are located at www.itk.org.

- Users and developers alike should read the ITK Software Guide

- Many compilable examples are available on the ITK Examples Wiki

- Tutorials are available at www.itk.org/ITK/help/tutorials.html

- The software can be downloaded from www.itk.org/ITK/resources/software.html.

- Developers, or users interested in contributing code, should look in the document Insight/Documentation/InsightDeveloperStart. or InsightDeveloperStart.doc found in the source code distribution.

- Developers should also look at the ITK style guide Insight/Documentation/Style.pdf found in the source distribution.

23.4.2 Applications

A great way to learn about ITK is to see how it is used. There are four places to find applications of ITK.

1. The Insight/Examples/ source code examples distributed with ITK. The source code is available. In addition, it is heavily commented and works in combination with the ITK Software Guide.

2. The separate InsightApplications checkout.

3. The Applications web pages. These are extensive descriptions, with images and references, of the examples found in #1 above.

4. The testing directories distributed with ITK are simple, mainly undocumented examples of how to use the code.

In 2004 ITK-SNAP (website) was developed from SNAP and became a popular free segmentation software using ITK and having a nice and simple user interface.

23.4.3 Data

- Data is available via anonymous ftp from: public.kitware.com/pub/itk/Data/.

- See also the ITK Data web page.

23.4.4 See also

23.4.5 Contacts

- Terry Yoo (NLM/NIH Insight Project Manager yoo at nlm.nih.gov)

- Will Schroeder (PI Kitware, Inc. will.schroeder at kitware.com)

23.5   References

[1] "Insight Software Consortium / ITK - GitHub".

[2] "Copyright and License".

- Yoo, TS; Ackerman, MJ; Lorensen, WE; et al. (2002). "Engineering and algorithm design for an image processing Api: a technical report on ITK—the Insight Toolkit". *Stud Health Technol Inform* **85**: 586–92. PMID 15458157.

- Yoo, TS; Metaxas, DN (Dec 2005). "Open science—combining open data and open source software: medical image analysis with the Insight Toolkit". *Med Image Anal* **9** (6): 503–6. doi:10.1016/j.media.2005.04.008. PMID 16169766.

- Prior, FW; Erickson, BJ; Tarbox, L (Nov 2007). "Open source software projects of the caBIG In Vivo Imaging Workspace Software special interest group". *J Digit Imaging* **20** (Suppl 1): 94–100. doi:10.1007/s10278-007-9061-4. PMC 2039820. PMID 17846835.

Chapter 24

Integrating Vision Toolkit

The **Integrating Vision Toolkit** (IVT) is a C++ computer vision library with an object-oriented architecture. It offers its own multi-platform GUI toolkit.

24.1   Availability

The library is available as free software under a 3-clause BSD license. It is written in pure ANSI C++ and compiles using any available C++ compiler (e.g. any Visual Studio, any gcc, TI Code Composer). It is cross-platform and runs on basically any platform offering a C++ compiler, including Windows, Mac OS X and Linux. The included GUI toolkit offers implementations for Windows (Win32 API), Linux (GTK), Mac OS X (Cocoa) and Qt. The computer vision company Keyetech offers platform specific optimizations of various IVT functions with the Keyetech Performance Primitives (KPP), which are automatically loaded by the IVT.

24.2   History

The IVT has been developed at the formerly named University of Karlsruhe (TH), now Karlsruhe Institute of Technology (KIT). The first version of the IVT was released on Sourceforge on December 22, 2005. Since 2009, the IVT is maintained in cooperation with the company Keyetech,[1] who also offers training courses for the IVT as well as commercial products and customized solutions using the IVT.

24.3   Features

IVT's features include:

- Camera interface and implementation for various cameras (IEEE1394 (CMU1394, Unicap1394), USB webcams (V4L and VfW), Quicktime, PointGrey Dragonfly, Videre SVS, etc.)

- Camera model (monocular and stereo)

- Camera calibration (using OpenCV 1.0)

- Filters (Gaussian Smoothing, Sobel, Prewitt, Canny)

- Color segmentation (HSV color space)

- Hough transform (lines and circles)

81

- Point operations (affine, thresholding)

- SIFT features

- Harris corner detector

- Stereo vision (disparity map computation, stereo triangulation)

- Undistortion

- Rectification

- Linear least squares

- Particle filtering framework

- Data structures (matrices, vectors, kd-tree, dynamic array)

- POSIT (6 DoF pose from 2D-3D point correspondences)

- Own GUI toolkit with implementations for several platforms (by Florian Hecht) (see below)

24.4   Relation to OpenCV

Compared to OpenCV the IVT offers an object-oriented software architecture, is easier to read and easier to use. The implementations are as fast or even faster than those from the OpenCV. However, OpenCV offers some functionality that IVT does not offer (e.g. a face detector). Such functionality is integrated by optional OpenCV wrappers.

24.5   See also

- VXL

24.6   References

[1] Keyetech

24.7   External links

- Integrating Vision Toolkit on SourceForge.net - (with documentation, installation guide, example applications, and links for support, services and related books)

Chapter 25

Intel RealSense

Intel RealSense, formerly known as **Intel Perceptual Computing**,[1] is a platform for implementing gesture-based Human-Computer Interaction techniques. It consists of series of consumer grade 3D cameras together with an easy to use machine perception library that simplifies supporting the cameras for third-party software developers.[2][3][4]

As of March 2015, multiple laptop and tablet computer manufactures offer one or more devices with Intel RealSense camera built in. These are Asus, HP, Dell, Lenovo, and Acer.[5]

25.1   Features

- Facial analysis
  - Tracking multiple faces.
  - Identification of facial features like eyes, mouth and nose
- Hand and finger tracking
  - Tracking up to 10 simultaneous fingers, 8 gestures, and access to raw depth data
- Sound processing
  - Speech recognition
  - Background noise subtraction
- Augmented reality
  - Object tracking
  - Drawing CG images on real world scenarios

25.2   3D cameras

An Intel RealSense camera contains the following four components. A conventional camera, an infrared laser projector, an infrared camera, and a microphone array. The infrared projector projects a grid onto the scene (in infrared light which is invisible to human eye) and the infrared camera records it to compute depth information. The microphone array allows localizing sound sources in space and performing background noise cancellation.

Three camera models were announced, with distinct specifications and intended use.

25.2.1 Intel RealSense 3D Camera (Front F200)

This is a stand-alone camera that can be attached to a desktop or laptop computer.[6] It is intended to be used for natural gesture-based interaction, face recognition, immersive, video conferencing and collaboration, gaming and learning and 3D scanning.[7]

There is a version of this camera to be embedded into laptop computers.[8]

Specifications

- Full VGA depth resolution

- 1080p RGB camera

- 0.2–1.2 meter range (Specific algorithms may have different range and accuracy)

- USB 3.0 interface

25.2.2 Intel RealSense Snapshot

Snapshot is a camera intended to be built into tablet computers and possibly smartphones. Its intended uses include taking photographs and performing after the fact refocusing, distance measurements, and applying motion photo filters.[9]

The refocus feature differs from a plenoptic camera in that RealSense Snapshot takes pictures with large depth of field so that initially the whole picture is in focus and then in software it selectively blurs parts of the image depending on their distance.

Dell Venue 8 7000 Series Android tablet is equipped with this camera.[10]

25.2.3 Intel RealSense 3D Camera (Rear R200)

Rear-mounted camera for Microsoft Surface or a similar tablet. It is not yet available on the market.[3] This camera is intended for augmented reality applications, content creation, and object scanning.

25.3 Developer kits

The Front 200 camera can be bought from Intel for approximately $100 USD.[11]

25.4 App Challenge

To address the lack of applications built on the RealSense platform and to promote the platform among software developers, in 2014 Intel organized the Intel RealSense App Challenge. The winners will be awarded large sums of money.[12]

25.5 Reception

In an early preview article, *PC World*'s Mark Hachman concluded that RealSense is an enabling technology that will be largely defined by the software that will take advantage of its features. He noted that as of the time the article was written, the technology was new and there was no such software.[3]

Intel dedicated large part of their presentation[13] at 2015 International Consumer Electronics Show in Las Vegas to RealSense.

25.6   See also

- Open CV

- Kinect

- Project Tango

- Creative Labs

25.7   References

[1] "ntel® RealSense™ SDK (former Intel® Perceptual Computing SDK) support (Bug #3917)". Retrieved 25 March 2015.

[2] . Intel https://software.intel.com/en-us/realsense/home. Missing or empty |title= (help)

[3] "Hands on: Without apps, Intel's RealSense camera is a puzzle". *PC World*. Mar 5, 2015. Retrieved 25 March 2015.

[4] . Intel https://software.intel.com/en-us/realsense/get-started. Missing or empty |title= (help)

[5] "Devices with Intel RealSense Technology". Intel.

[6] . Intel http://www.intel.com/content/www/us/en/architecture-and-technology/realsense-3d-camera.html. Missing or empty |title= (help)

[7] . Intel https://software.intel.com/en-us/RealSense/F200Camera. Missing or empty |title= (help)

[8] . Intel http://www.intel.com/content/www/us/en/architecture-and-technology/realsense-devices.html. Missing or empty |title= (help)

[9] . Intel https://software.intel.com/en-us/RealSense/Snapshot. Missing or empty |title= (help)

[10] . Intel http://www.intel.com/content/www/us/en/architecture-and-technology/realsense-snapshot.html. Missing or empty |title= (help)

[11] . Intel https://software.intel.com/en-us/RealSense/Devkit. Missing or empty |title= (help)

[12] "Intel RealSense App Challenge". Intel.

[13] . International CES https://www.youtube.com/watch?v=x6a3tKOwN7A. Missing or empty |title= (help)

25.8   External links

- Intel RealSense SDK developer documentation

Chapter 26

MATLAB

For the region in Bangladesh, see Matlab (Bangladesh).
Not to be confused with MATHLAB.

MATLAB (**mat**rix **lab**oratory) is a multi-paradigm numerical computing environment and fourth-generation programming language. A proprietary programming language developed by MathWorks, MATLAB allows matrix manipulations, plotting of functions and data, implementation of algorithms, creation of user interfaces, and interfacing with programs written in other languages, including C, C++, Java, Fortran and Python.

Although MATLAB is intended primarily for numerical computing, an optional toolbox uses the MuPAD symbolic engine, allowing access to symbolic computing capabilities. An additional package, Simulink, adds graphical multi-domain simulation and model-based design for dynamic and embedded systems.

In 2004, MATLAB had around one million users across industry and academia.[3] MATLAB users come from various backgrounds of engineering, science, and economics.

26.1 History

Cleve Moler, the chairman of the computer science department at the University of New Mexico, started developing MATLAB in the late 1970s.[4] He designed it to give his students access to LINPACK and EISPACK without them having to learn Fortran. It soon spread to other universities and found a strong audience within the applied mathematics community. Jack Little, an engineer, was exposed to it during a visit Moler made to Stanford University in 1983. Recognizing its commercial potential, he joined with Moler and Steve Bangert. They rewrote MATLAB in C and founded MathWorks in 1984 to continue its development. These rewritten libraries were known as JACKPAC.[5] In 2000, MATLAB was rewritten to use a newer set of libraries for matrix manipulation, LAPACK.[6]

MATLAB was first adopted by researchers and practitioners in control engineering, Little's specialty, but quickly spread to many other domains. It is now also used in education, in particular the teaching of linear algebra, numerical analysis, and is popular amongst scientists involved in image processing.[4]

26.2 Syntax

The MATLAB application is built around the MATLAB scripting language. Common usage of the MATLAB application involves using the Command Window as an interactive mathematical shell or executing text files containing MATLAB code.[7]

26.2.1 Variables

Variables are defined using the assignment operator, =. MATLAB is a weakly typed programming language because types are implicitly converted.[8] It is an inferred typed language because variables can be assigned without declaring their type, except if they are to be treated as symbolic objects,[9] and that their type can change. Values can come from constants, from computation involving values of other variables, or from the output of a function. For example:

>> x = 17 x = 17 >> x = 'hat' x = hat >> y = x + 0 y = 104 97 116 >> x = [3*4, pi/2] x = 12.0000 1.5708 >> y = 3*sin(x) y = −1.6097 3.0000

26.2.2 Vectors and matrices

A simple array is defined using the colon syntax: *init:increment:terminator*. For instance:

>> array = 1:2:9 array = 1 3 5 7 9

defines a variable named array (or assigns a new value to an existing variable with the name array) which is an array consisting of the values 1, 3, 5, 7, and 9. That is, the array starts at 1 (the *init* value), increments with each step from the previous value by 2 (the *increment* value), and stops once it reaches (or to avoid exceeding) 9 (the *terminator* value).

>> array = 1:3:9 array = 1 4 7

the *increment* value can actually be left out of this syntax (along with one of the colons), to use a default value of 1.

>> ari = 1:5 ari = 1 2 3 4 5

assigns to the variable named ari an array with the values 1, 2, 3, 4, and 5, since the default value of 1 is used as the incrementer.

Indexing is one-based,[10] which is the usual convention for matrices in mathematics, although not for some programming languages such as C, C++, and Java.

Matrices can be defined by separating the elements of a row with blank space or comma and using a semicolon to terminate each row. The list of elements should be surrounded by square brackets: []. Parentheses: () are used to access elements and subarrays (they are also used to denote a function argument list).

>> A = [16 3 2 13; 5 10 11 8; 9 6 7 12; 4 15 14 1] A = 16 3 2 13 5 10 11 8 9 6 7 12 4 15 14 1 >> A(2,3) ans = 11

Sets of indices can be specified by expressions such as "2:4", which evaluates to [2, 3, 4]. For example, a submatrix taken from rows 2 through 4 and columns 3 through 4 can be written as:

>> A(2:4,3:4) ans = 11 8 7 12 14 1

A square identity matrix of size *n* can be generated using the function *eye*, and matrices of any size with zeros or ones can be generated with the functions *zeros* and *ones*, respectively.

>> eye(3,3) ans = 1 0 0 0 1 0 0 0 1 >> zeros(2,3) ans = 0 0 0 0 0 0 >> ones(2,3) ans = 1 1 1 1 1 1

Most MATLAB functions can accept matrices and will apply themselves to each element. For example, mod(2*J,n) will multiply every element in "J" by 2, and then reduce each element modulo "n". MATLAB does include standard "for" and "while" loops, but (as in other similar applications such as R), using the vectorized notation often produces code that is faster to execute. This code, excerpted from the function *magic.m*, creates a magic square *M* for odd values of *n* (MATLAB function meshgrid is used here to generate square matrices I and J containing 1:n).

[J,I] = meshgrid(1:n); A = mod(I + J - (n + 3) / 2, n); B = mod(I + 2 * J - 2, n); M = n * A + B + 1;

26.2.3   Structures

MATLAB has structure data types.[11] Since all variables in MATLAB are arrays, a more adequate name is "structure array", where each element of the array has the same field names. In addition, MATLAB supports dynamic field names[12] (field look-ups by name, field manipulations, etc.). Unfortunately, MATLAB JIT does not support MATLAB structures, therefore just a simple bundling of various variables into a structure will come at a cost.[13]

26.2.4   Functions

When creating a MATLAB function, the name of the file should match the name of the first function in the file. Valid function names begin with an alphabetic character, and can contain letters, numbers, or underscores.

26.2.5   Function handles

MATLAB supports elements of lambda calculus by introducing function handles,[14] or function references, which are implemented either in .m files or anonymous[15]/nested functions.[16]

26.2.6   Classes and object-oriented programming

MATLAB's support for object-oriented programming includes classes, inheritance, virtual dispatch, packages, pass-by-value semantics, and pass-by-reference semantics.[17] However, the syntax and calling conventions are significantly different from other languages. MATLAB has value classes and reference classes, depending on whether the class has *handle* as a super-class (for reference classes) or not (for value classes).[18]

Method call behavior is different between value and reference classes. For example, a call to a method

object.method();

can alter any member of *object* only if *object* is an instance of a reference class.

An example of a simple class is provided below.

classdef hello methods function greet(this) disp('Hello!') end end end

When put into a file named hello.m, this can be executed with the following commands:

>> x = hello; >> x.greet(); Hello!

26.3   Graphics and graphical user interface programming

MATLAB supports developing applications with graphical user interface features. MATLAB includes GUIDE[19] (GUI development environment) for graphically designing GUIs.[20] It also has tightly integrated graph-plotting features. For example, the function *plot* can be used to produce a graph from two vectors x and y. The code:

x = 0:pi/100:2*pi; y = sin(x); plot(x,y)

produces the following figure of the sine function:

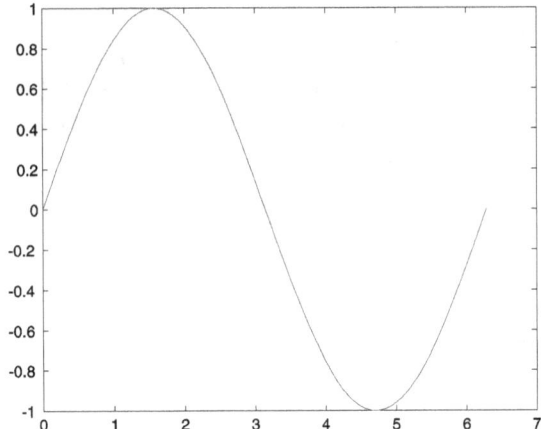

A MATLAB program can produce three-dimensional graphics using the functions *surf*, *plot3* or *mesh*.

In MATLAB, graphical user interfaces can be programmed with the GUI design environment (GUIDE) tool.[21]

26.4 Interfacing with other languages

MATLAB can call functions and subroutines written in the C programming language or Fortran.[22] A wrapper function is created allowing MATLAB data types to be passed and returned. The dynamically loadable object files created by compiling such functions are termed "MEX-files" (for **MATLAB ex**ecutable).[23][24] Since 2014 increasing two-way interfacing with Python is being added.[25][26]

Libraries written in Perl, Java, ActiveX or .NET can be directly called from MATLAB,[27][28] and many MATLAB libraries (for example XML or SQL support) are implemented as wrappers around Java or ActiveX libraries. Calling MATLAB from Java is more complicated, but can be done with a MATLAB toolbox[29] which is sold separately by MathWorks, or using an undocumented mechanism called JMI (Java-to-MATLAB Interface),[30][31] (which should not be confused with the unrelated Java Metadata Interface that is also called JMI).

As alternatives to the MuPAD based Symbolic Math Toolbox available from MathWorks, MATLAB can be connected to Maple or Mathematica.[32][33]

Libraries also exist to import and export MathML.[34]

26.5 License

MATLAB is a proprietary product of MathWorks, so users are subject to vendor lock-in.[3][35] Although MATLAB Builder products can deploy MATLAB functions as library files which can be used with .NET[36] or Java[37] application building environment, future development will still be tied to the MATLAB language.

Each toolbox is purchased separately. If an evaluation license is requested, the MathWorks sales department requires detailed information about the project for which MATLAB is to be evaluated. If granted (which it often is), the evaluation license is valid for two to four weeks. A student version of MATLAB is available as is a home-use license for MATLAB, SIMULINK, and a subset of Mathwork's Toolboxes at substantially reduced prices.

It has been reported that EU competition regulators are investigating whether MathWorks refused to sell licenses to a competitor.[38] The regulators dropped the investigation after the complainant withdrew their accusation and no evidence of wrongdoing was found.[39]

26.6   Alternatives

See also: list of numerical analysis software and comparison of numerical analysis software

MATLAB has a number of competitors.[40] Commercial competitors include Mathematica, TK Solver, Maple, and IDL. There are also free open source alternatives to MATLAB, in particular GNU Octave, Scilab, FreeMat, Julia, and Sage which are intended to be mostly compatible with the MATLAB language. Among other languages that treat arrays as basic entities (array programming languages) are APL, Fortran 90 and higher, S-Lang, as well as the statistical languages R and S. There are also libraries to add similar functionality to existing languages, such as IT++ for C++, Perl Data Language for Perl, ILNumerics for .NET, NumPy/SciPy for Python, and Numeric.js for JavaScript.

GNU Octave is unique from other alternatives because it treats incompatibility with MATLAB as a bug (see MATLAB Compatibility of GNU Octave). Therefore, GNU Octave attempts to provide a software clone of MATLAB.

26.7   Release history

The number (or Release number) is the version reported by Concurrent License Manager program FLEXlm.

For a complete list of changes of both MATLAB and official toolboxes, consult the MATLAB release notes.[74]

26.8   File extensions

26.8.1   MATLAB

.fig MATLAB figure

.m MATLAB code (function, script, or class)

.mat MATLAB data (binary file for storing variables)

.mex... (.mexw32, .mexw64, .mexglx, ...) MATLAB executable MEX-files[75] (platform specific, e.g. ".mexmac" for the Mac, ".mexglx" for Linux, etc.[76])

.p MATLAB content-obscured .m file (P-code[77])

.mlappinstall MATLAB packaged App Installer[78]

.mlpkginstall support package installer (add-on for third-party hardware)[79]

.mltbx packaged custom toolbox[80]

.prj project file used by various solutions (packaged app/toolbox projects, MATLAB Compiler/Coder projects, Simulink projects)

.rpt report setup file created by MATLAB Report Generator[81]

26.8.2   Simulink

.mdl Simulink Model

.mdlp Simulink Protected Model

.slx Simulink Model (SLX format)

.slxp Simulink Protected Model (SLX format)

26.8.3 Simscape

.ssc Simscape[82] Model

26.8.4 MuPAD

.mn MuPAD Notebook

.mu MuPAD Code

.xvc, .xvz MuPAD Graphics

26.8.5 Third-party

.jkt GPU Cache file generated by Jacket for MATLAB (AccelerEyes)

.mum MATLAB CAPE-OPEN Unit Operation Model File (AmsterCHEM)

26.9 Easter eggs

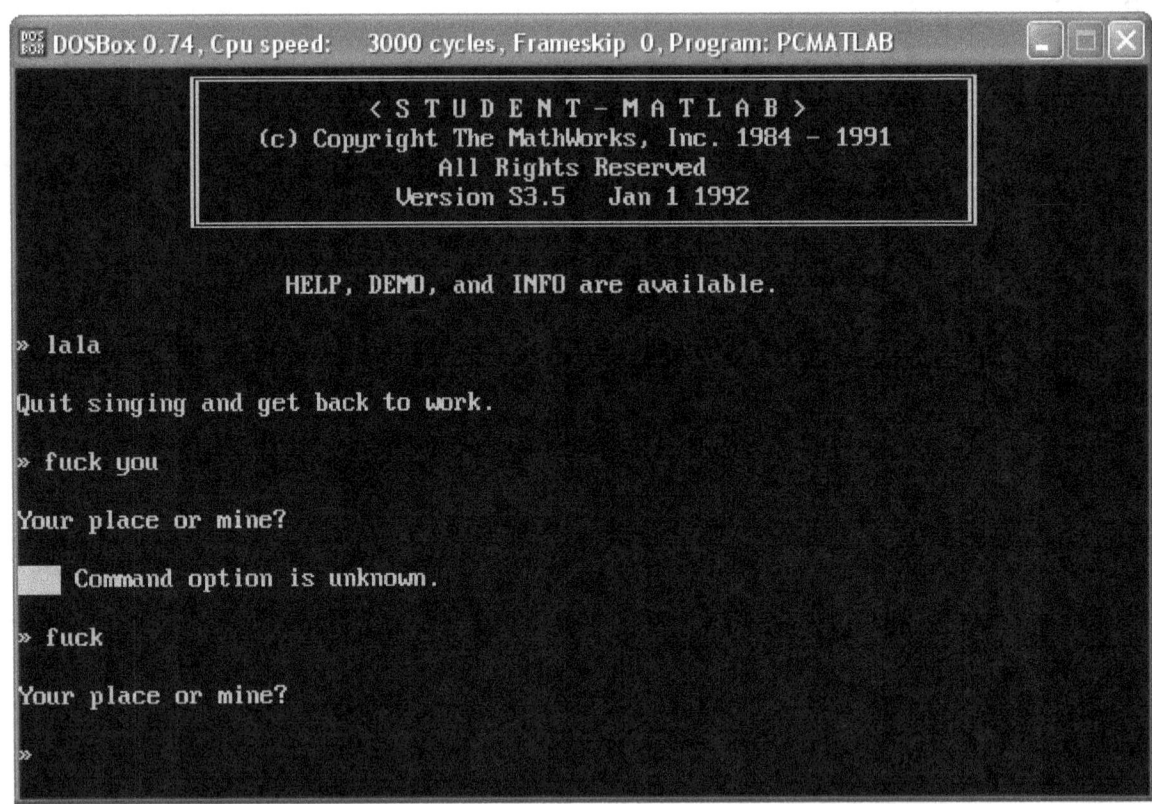

Screen capture of two easter eggs in MATLAB 3.5.

Several easter eggs exist in MATLAB.[83] These include hidden pictures,[84] and jokes. For example, typing in "spy" will generate a picture of the spies from Spy vs Spy. "Spy" was changed to an image of a dog in recent releases (R2011B). Typing in "why" randomly outputs a philosophical answer. Other commands include "penny", "toilet", "image", and "life". Not every Easter egg appears in every version of MATLAB.

26.10   See also

- Comparison of numerical analysis software

- List of numerical analysis software

26.11   Notes

[1] "The L-Shaped Membrane". MathWorks. 2003. Retrieved 7 February 2014.

[2] "System Requirements and Platform Availability". MathWorks. Retrieved 14 August 2013.

[3] Goering, Richard (4 October 2004). "Matlab edges closer to electronic design automation world". *EE Times*.

[4] Cleve Moler (December 2004). "The Origins of MATLAB". Retrieved 15 April 2007.

[5] "MATLAB Programming Language". Altius Directory. Retrieved 17 December 2010.

[6] Moler, Cleve (January 2000). "MATLAB Incorporates LAPACK". *Cleve's Corner*. MathWorks. Retrieved 20 December 2008.

[7] "MATLAB Documentation". MathWorks. Retrieved 14 August 2013.

[8] "Comparing MATLAB with Other OO Languages". *MATLAB*. MathWorks. Retrieved 14 August 2013.

[9] "Create Symbolic Variables and Expressions". *Symbolic Math Toolbox*. MathWorks. Retrieved 14 August 2013.

[10] "Matrix Indexing". MathWorks. Retrieved 14 August 2013.

[11] "Structures". MathWorks. Retrieved 14 August 2013.

[12] "Generate Field Names from Variables". MathWorks. Retrieved 14 August 2013.

[13] Considering Performance in Object-Oriented MATLAB Code, Loren Shure, MATLAB Central, 26 March 2012: "function calls on structs, cells, and function handles will not benefit from JIT optimization of the function call and can be many times slower than function calls on purely numeric arguments"

[14] "Function Handles". MathWorks. Retrieved 14 August 2013.

[15] "Anonymous Functions". MathWorks. Retrieved 14 August 2013.

[16] "Nested Functions". MathWorks.

[17] "Object-Oriented Programming". MathWorks. Retrieved 14 August 2013.

[18] "Comparing Handle and Value Classes". MathWorks.

[19] "Create a Simple GUIDE GUI". MathWorks. Retrieved 14 August 2014.

[20] "MATLAB GUI". MathWorks. 30 April 2011. Retrieved 14 August 2013.

[21] Smith, S. T. (2006). *Matlab: Advanced GUI Development*. Dog Ear Publishing. ISBN 978-1-59858-181-2.

[22] "Application Programming Interfaces to MATLAB". MathWorks. Retrieved 14 August 2013.

[23] "Create MEX-Files". MathWorks. Retrieved 14 August 2013.

[24] Spielman, Dan (10 February 2004). "Connecting C and Matlab". Yale University, Computer Science Department. Retrieved 20 May 2008.

[25] "MATLAB Engine for Python". MathWorks. Retrieved 13 June 2015.

[26] "Call Python Libraries". MathWorks. Retrieved 13 June 2015.

[27] "External Programming Language Interfaces". MathWorks. Retrieved 14 August 2013.

[28] "Call Perl script using appropriate operating system executable". MathWorks. Retrieved 7 November 2013.

[29] "MATLAB Builder JA". MathWorks. Retrieved 7 June 2010.

[30] Altman, Yair (14 April 2010). "Java-to-Matlab Interface". Undocumented Matlab. Retrieved 7 June 2010.

[31] Kaplan, Joshua. "matlabcontrol JMI".

[32] Germundsson, Roger (30 September 1998). "MaMa: Calling MATLAB from Mathematica with MathLink". *Wolfram Research*. Wolfram Library Archive.

[33] rsmenon; szhorvat (2013). "MATLink: Communicate with MATLAB from Mathematica". Retrieved 14 August 2013.

[34] Weitzel, Michael (1 September 2006). "MathML import/export". MathWorks - File Exchange. Retrieved 14 August 2013.

[35] Stafford, Jan (21 May 2003). "The Wrong Choice: Locked in by license restrictions". SearchOpenSource.com. Retrieved 14 August 2013.

[36] "MATLAB Builder NE". MathWorks. Retrieved 14 August 2013.

[37] "MATLAB Builder JA". MathWorks. Retrieved 14 August 2013.

[38] "MathWorks Software Licenses Probed by EU Antitrust Regulators". Bloomberg news. 1 March 2012.

[39] "EU regulators scrap antitrust case against MathWorks". Reuters. 2 Sep 2014.

[40] Steinhaus, Stefan (24 February 2008). "Comparison of mathematical programs for data analysis".

[41] Moler, Cleve (January 2006). "The Growth of MATLAB and The MathWorks over Two Decades". *News & Notes Newsletter*. MathWorks. Retrieved 14 August 2013.

[42] "MATLAB System Requirements - Release 13". MathWorks. Retrieved 6 October 2015.

[43] "Memory Mapping". MathWorks. Retrieved 22 January 2014.

[44] "MATLAB bsxfun". MathWorks. Retrieved 22 January 2014.

[45] "Do MATLAB versions prior to R2007a run under Windows Vista?". MathWorks. 3 September 2010. Retrieved 8 February 2011.

[46] "OOP Compatibility with Previous Versions". MathWorks. Retrieved 11 March 2013.

[47] "Packages Create Namespaces". MathWorks. Retrieved 22 January 2014.

[48] "Map Containers". MathWorks. Retrieved 22 January 2014.

[49] "Creating and Controlling a Random Number Stream". MathWorks. Retrieved 22 January 2014.

[50] "New MATLAB External Interfacing Features in R2009a". MathWorks. Retrieved 22 January 2014.

[51] "Ignore Function Outputs". MathWorks. Retrieved 22 January 2014.

[52] "Ignore Function Inputs". MathWorks. Retrieved 22 January 2014.

[53] "Working with Enumerations". MathWorks. Retrieved 22 January 2014.

[54] "What's New in Release 2010b". MathWorks. Retrieved 22 January 2014.

[55] "New RNG Function for Controlling Random Number Generation in Release 2011a". MathWorks. Retrieved 22 January 2014.

[56] "MATLAB rng". MathWorks. Retrieved 22 January 2014.

[57] "Replace Discouraged Syntaxes of rand and randn". MathWorks. Retrieved 22 January 2014.

[58] "MATLAB matfile". MathWorks. Retrieved 22 January 2014.

[59] "MATLAB max workers". Retrieved 22 January 2014.

[60] Shure, Loren (September 2012). "The MATLAB R2012b Desktop – Part 1: Introduction to the Toolstrip".

[61] "MATLAB Apps". MathWorks. Retrieved 14 August 2013.

[62] "MATLAB Unit Testing Framework". MathWorks. Retrieved 14 August 2013.

[63] "MathWorks Announces Release 2013b of the MATLAB and Simulink Product Families". MathWorks. September 2013.

[64] "MATLAB Tables". MathWorks. Retrieved 14 September 2013.

[65] "MathWorks Announces Release 2014a of the MATLAB and Simulink Product Families". MathWorks. Retrieved 11 March 2014.

[66] "Graphics Changes in R2014b". MathWorks. Retrieved 3 October 2014.

[67] "uitab: Create tabbed panel". MathWorks. Retrieved 3 October 2014.

[68] "Create and Share Toolboxes". MathWorks. Retrieved 3 October 2014.

[69] "Dates and Time". MathWorks. Retrieved 3 October 2014.

[70] "Source Control Integration". MathWorks. Retrieved 3 October 2014.

[71] "MATLAB MapReduce and Hadoop". MathWorks. Retrieved 3 October 2014.

[72] "Call Python Libraries". MathWorks. Retrieved 3 October 2014.

[73] "MATLAB Engine for Python". MathWorks. Retrieved 3 October 2014.

[74] "MATLAB Release Notes". MathWorks. Retrieved 25 January 2014.

[75] "Introducing MEX-Files". MathWorks. Retrieved 14 August 2013.

[76] "Binary MEX-File Extensions". MathWorks. Retrieved 14 August 2013.

[77] "Protect Your Source Code". MathWorks. Retrieved 14 August 2013.

[78] "MATLAB App Installer File". MathWorks. Retrieved 14 August 2013.

[79] "Support Package Installation". MathWorks. Retrieved 3 October 2014.

[80] "Manage Toolboxes". MathWorks. Retrieved 3 October 2014.

[81] "MATLAB Report Generator". MathWorks. Retrieved 3 October 2014.

[82] "Simscape". MathWorks. Retrieved 14 August 2013.

[83] "What MATLAB Easter eggs do you know?". MathWorks - MATLAB Answers. 25 February 2011. Retrieved 14 August 2013.

[84] Eddins, Steve (17 October 2006). "The Story Behind the MATLAB Default Image". Retrieved 14 August 2013.

26.12   References

- Gilat, Amos (2004). *MATLAB: An Introduction with Applications 2nd Edition*. John Wiley & Sons. ISBN 978-0-471-69420-5.

- Quarteroni, Alfio; Saleri, Fausto (2006). *Scientific Computing with MATLAB and Octave*. Springer. ISBN 978-3-540-32612-0.

- Ferreira, A.J.M. (2009). *MATLAB Codes for Finite Element Analysis*. Springer. ISBN 978-1-4020-9199-5.

- Lynch, Stephen (2004). *Dynamical Systems with Applications using MATLAB*. Birkhäuser. ISBN 978-0-8176-4321-8.

26.13   External links

- Official website

- MATLAB Central File Exchange – Library of over 20,000 user-contributed MATLAB files and toolboxes, mostly distributed under BSD License.

- MATLAB at DMOZ

- MATLAB Central Newsreader – a web-based newsgroups reader hosted by MathWorks for comp.soft-sys.matlab

- LiteratePrograms (MATLAB)

- MATLAB Central Blogs

- *Physical Modeling in MATLAB* by Allen B. Downey, Green Tea Press, PDF, ISBN 978-0-615-18550-7. An introduction to MATLAB.

- *Writing Fast MATLAB Code* by Pascal Getreuer

- Calling MATLAB from Java: MatlabControl JMI Wrapper, The MatlabJava Server, MatlabControl

- International Online Workshop on MATLAB and Simulink by WorldServe Education

- MATLAB tag on Stack Overflow.

- MATLAB Answers – a collaborative environment for finding the best answers to your questions about MATLAB, Simulink, and related products.

- Cody – a MATLAB Central game that challenges and expands your knowledge of MATLAB.

- MATLAB Online Programming Contest

- Trendy – a MATLAB based web service for tracking and plotting trends.

- Undocumented Matlab – a blog on undocumented/non-official aspects of MATLAB.

- Hazewinkel, Michiel, ed. (2001), "Linear algebra software packages", *Encyclopedia of Mathematics*, Springer, ISBN 978-1-55608-010-4

- MATLAB free course on Wikiversity

Chapter 27

MeVisLab

MeVisLab is a cross-platform application framework for medical image processing and scientific visualization. It includes advanced algorithms for image registration, segmentation, and quantitative morphological and functional image analysis. An IDE for graphical programming and rapid user interface prototyping is available.

MeVisLab is written in C++ and uses the Qt framework for graphical user interfaces. It is available cross-platform on Windows, Linux, and Mac OS X. The software development is done in cooperation between MeVis Medical Solutions AG and Fraunhofer MEVIS.

A freeware version of the MeVislab SDK is available (see Licensing). Open source modules are delivered as MeVisLab Public Sources in the SDK and available from the MeVisLab Community and Community Sources project.

27.1 History

MeVisLab development began in 1993 with the software ILAB1 of the CeVis Institute, written in C++. It allowed to interactively connect algorithms of the Image Vision Library (IL) on Silicon Graphics (SGI) to form image processing networks. In 1995, the newly founded MeVis Research GmbH (which became Fraunhofer MEVIS in 2009) took over the ILAB development and released ILAB2 and ILAB3. OpenInventor and Tcl scripting was integrated but both programs were still running on SGI only.[1]

In 2000, ILAB4 was released with the core rewritten in Objective-C for Windows. For being able to move away from the SGI platform, the Image Vision Library was substituted by the platform-independent, inhouse-developed MeVis Image Processing Library (ML). In 2002, the code was adapted to work on the application framework Qt.[1]

In 2004, the software was released under the name MeVisLab. It contained an improved IDE and was available on Windows and Linux.[2] See the Release history for details.

In 2007, MeVisLab has been acquired by MeVis Medical Solutions AG. Since then, MeVisLab has been continued as a collaborative project between the MeVis Medical Solutions and Fraunhofer MEVIS.

27.2 Features

MeVisLab features include:[3][4][5]

- **Image processing with the MeVis Image Processing Library (ML)**: The ML is a request-driven, page-based, modular, expandable C++ image processing library supporting up to six image dimensions (x, y, z, color, time, user dimensions). It offers a priority-controlled page cache and high performance for large data sets.

- **2D image viewing**: Fast, modular, extensible 2D viewers with combined 2D/3D rendering are implemented, supporting slab rendering (volume rendering/MIP), overlays, point/ROI selection, Multiplanar Reformations (MPR),

96

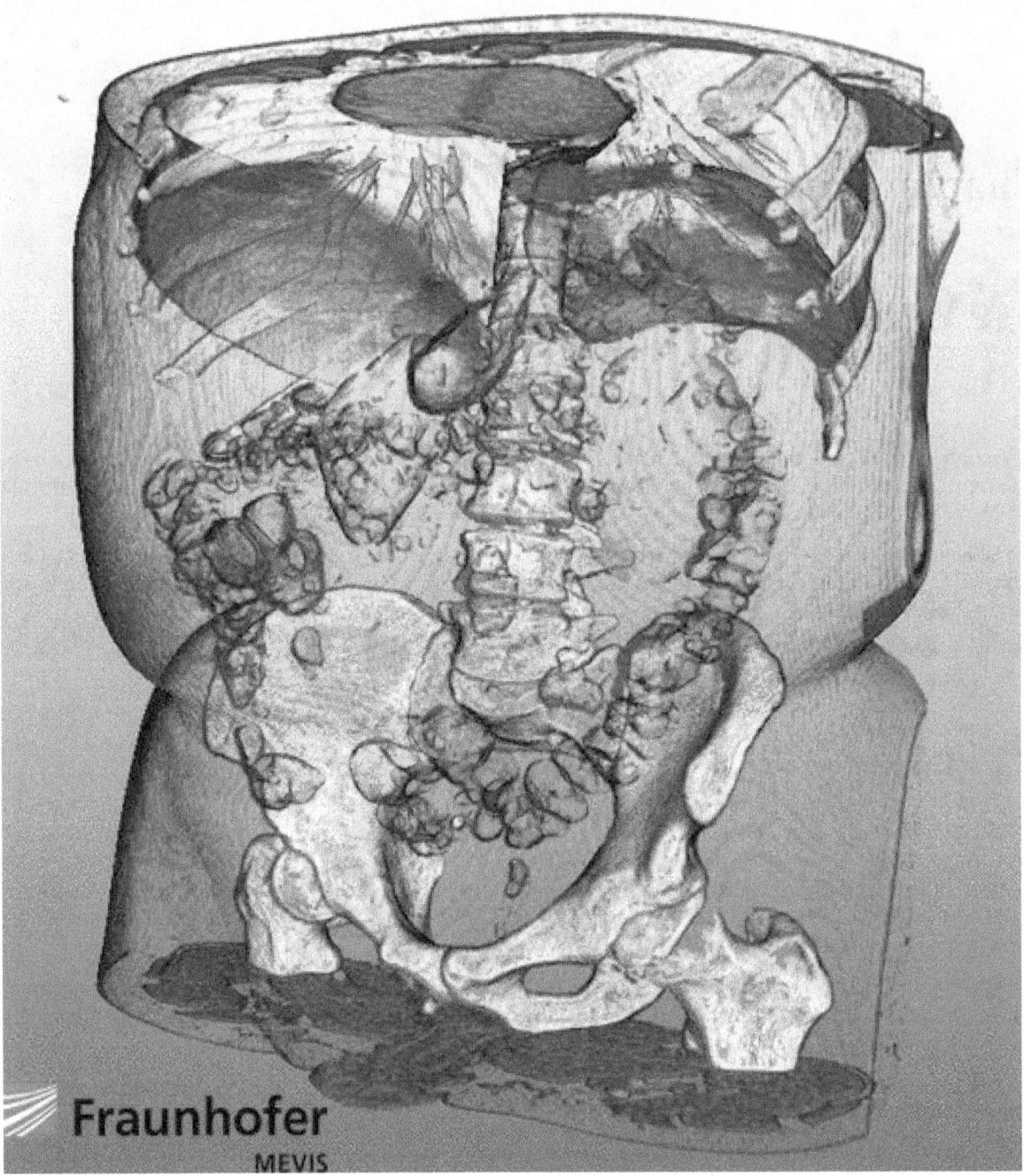

Body center rendered in MeVisLab

as well as interactive editing of marker objects (points, vectors, discs, spheres, etc.)

- **Volume rendering**: A high-quality volume renderer (Giga Voxel Renderer, GVR) based on OpenGL/Open Inventor is available.[6] It supports large image volumes (e.g., 512x512x2000 CT volumes, 12bit), time-varying data (e.g. dynamic MRI volumes), lookup tables, interactive region of interest, sub-volume selection, modular, multi-purpose GLSL shader framework.[7]

- **DICOM and other file formats**: DICOM is supported via an import step that automatically recognizes series of 2D DICOM frames that belong to the same 3D/4D image volume. The data can be browsed with a configurable

DICOM browser. DICOM storage to PACS is possible. Other supported file formats include TIFF (2D/3D, RGBA), Analyze, RAW, PNG, JPG, BMP, and more.

- **Tool frameworks**: Modular class and module libraries for markers, curves, histograms, Winged-Edged Meshes (WEM) and Contour Segmentation Objects (CSO) are available.

- **Qt integration**: Qt is used as application framework. The Qt API is integrated via PythonQt, allow to access Qt Style Sheets, Qt Widgets, QT Core classes, etc. by scripting from within MeVisLab.

- **Scripting support**: Python can be used for script controlled access to a large part of the MeVisLab functionality. The script binding to Qt is implemented via PythonQt. For image processing via Python, NumPy is available. Object-oriented Python programming in MeVisLab is possible.[8] JavaScript based on QSA is available as legacy support (QSA has been discontinued by Trolltech in 2008 in favor of QtScript).

- **Integrated open source image processing and visualization libraries**: Three open source libraries are integrated: Open Inventor, based on the original SGI source code released as open source in 2000;[9] Insight Toolkit (ITK), made available as MeVisLab modules;[10][11][12] Visualization Toolkit (VTK): made available as MeVisLab modules.[13][14]

- **Comprehensive module library**: The MeVisLab module library comprises a total of 2600 modules, including 800 standard modules and 1800 ITK/VTK modules.

27.3   MeVisLab principles

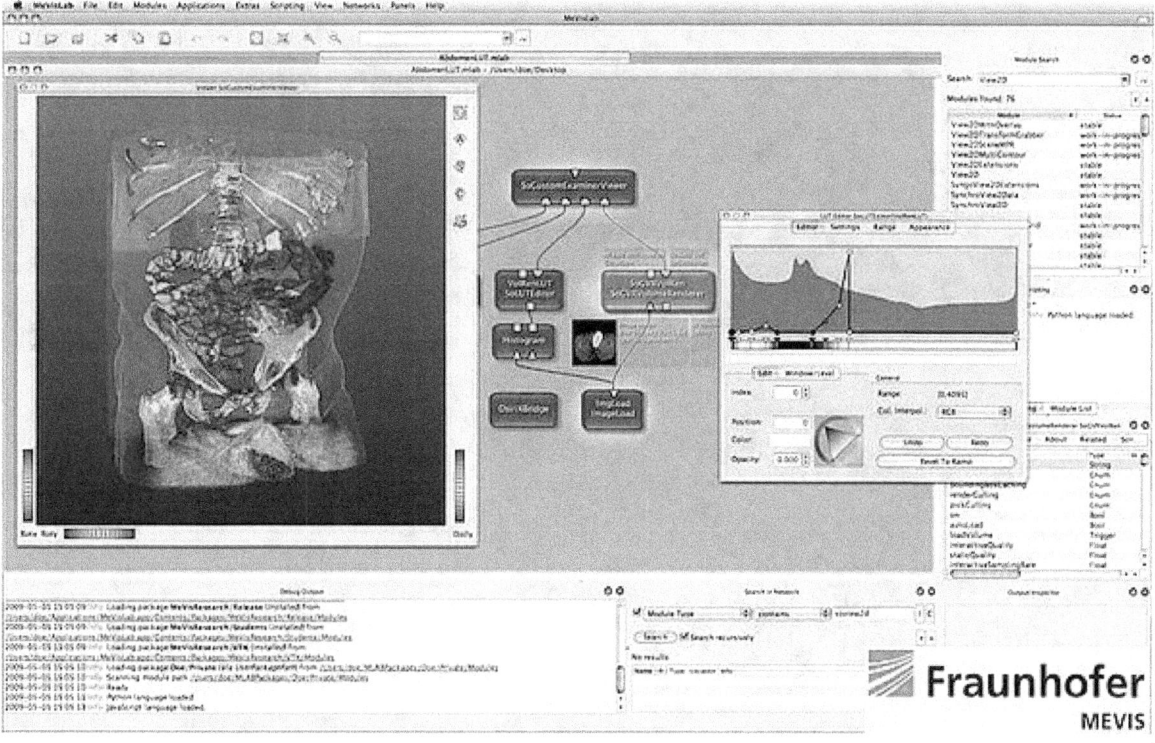

MeVisLab GUI

MeVisLab is a modular development framework. Based on modules, networks can be created and applications can be built.

To support the creation of image processing networks, MeVisLab offers an IDE that allows data-flow modelling by visual programming. Important IDE features are the multiple document interface (MDI), module and connection inspectors with docking ability, advanced search, scripting and debugging consoles, movie and screenshot generation and galleries, module testing and error handling support.[15]

In the visual network editor, modules can be added and combined to set up data flow and parameter synchronization. The resulting networks can be modified dynamically by scripts at runtime. Macro modules can be created to encapsulate subnetworks of modules, scripting functionality and high-level algorithms.

On top of the networks, the medical application level with viewers and UI panels can be added. Panels are written in the MeVisLab Definition Language (MDL), can be scripted with Python or JavaScript and styled using MeVisLab-internal mechanisms or Qt features.

The development of own modules written in C++ or Python is supported by wizards.

27.4   Image gallery

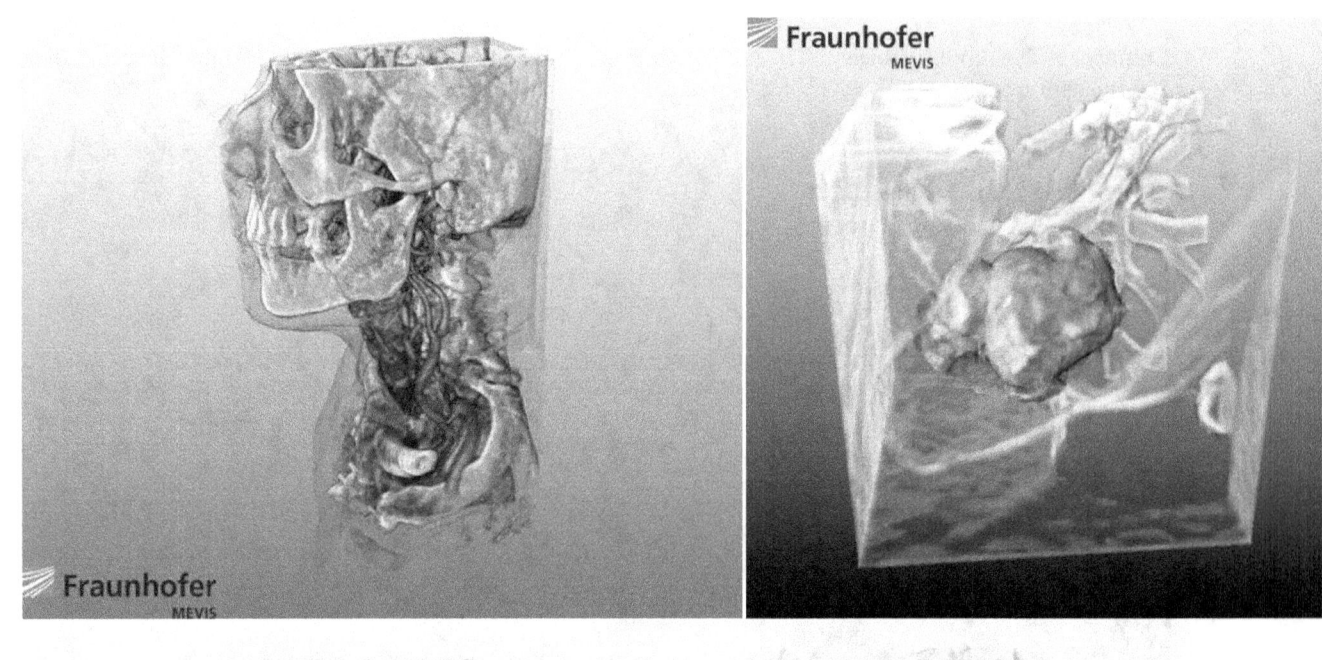

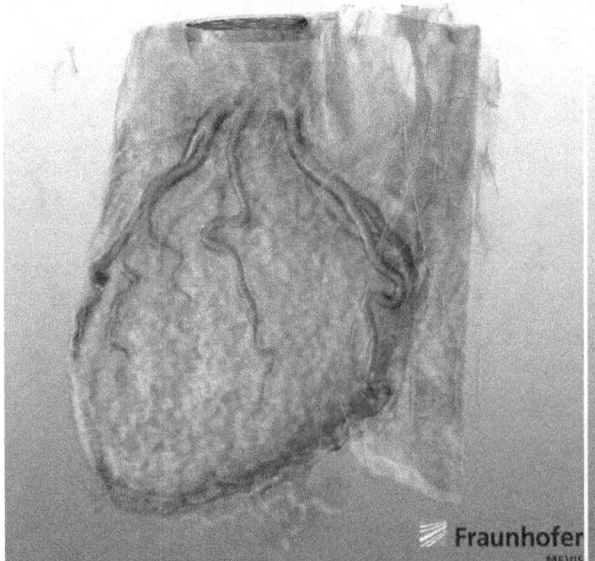

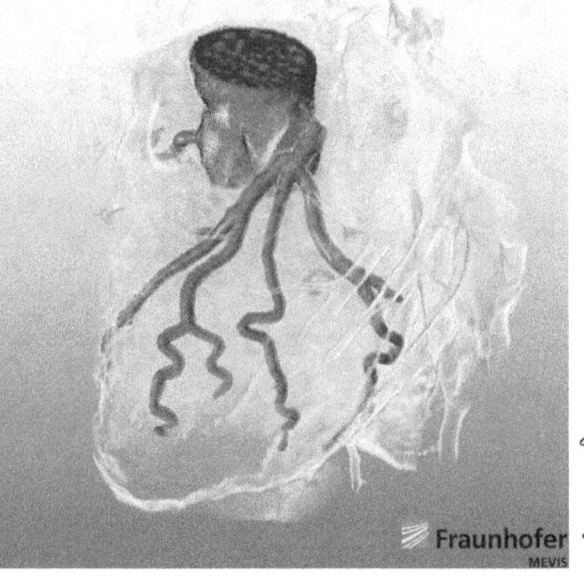

27.5   MeVisLab forum

MeVisLab offers a very well-supported public forum in which core developers as well as users of all levels of experience share information. A free registration is necessary.

27.6   Release history

The table below lists all main releases, without release candidates and maintenance releases. Various larger changes were made from version 1.6 to version 2.0. For detailed changes in the ML, see the ML Release Notes. For release news, see Release News on the MeVisLab Homepage.

27.7   Fields of application, research projects

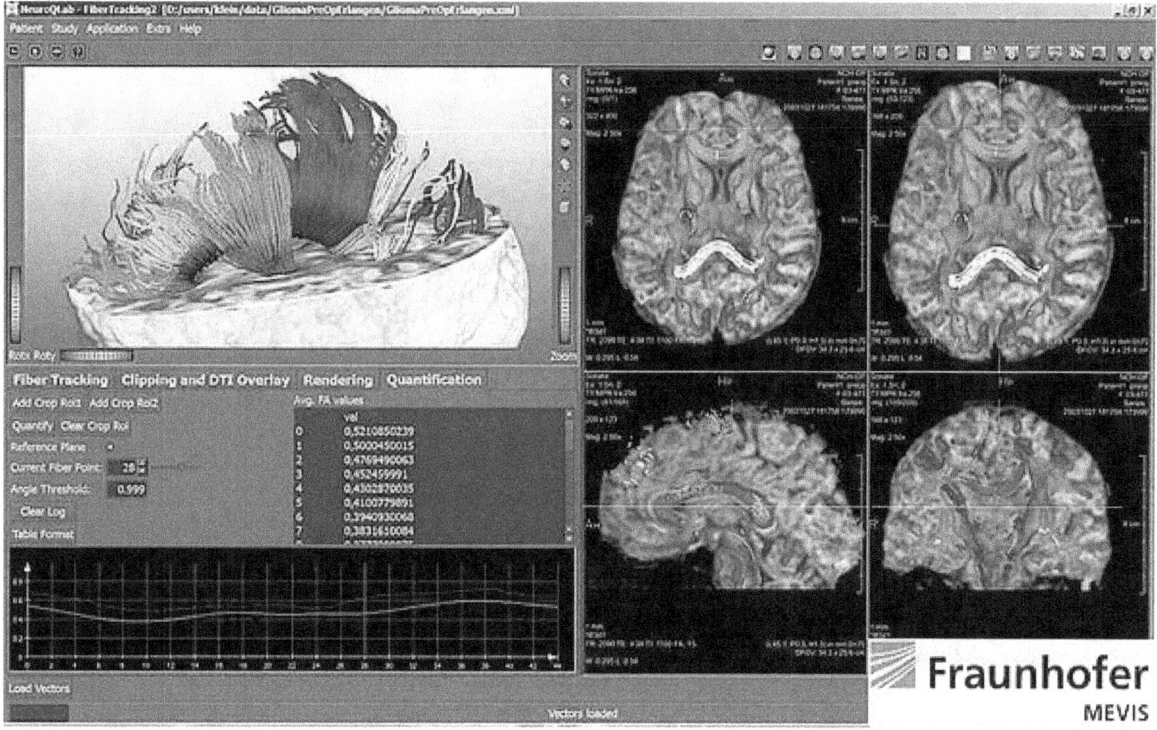

Application building with MeVisLab

MeVisLab has been used in a wide range of medical and clinical applications, including surgery planning[32] for liver,[33][34][35][36] lung,[37][38] head[39][40] and neck and other body regions, analysis of dynamic, contrast enhanced breast[41][42] and Prostate MRI, quantitative analysis of neurologic[43] and cardiovascular image series,[44][45] orthopedic quantification and visualization, tumor lesion volumetry[46] and therapy monitoring,[47] enhanced visualization of mammograms, 3D breast ultrasound and tomosynthesis image data, and many other applications. MeVisLab is also used as a training and teaching tool[48][49] for image processing (both general and medical[50]) and visualization techniques.

MeVisLab is and has been used in many research projects, including:

- VICORA VICORA Virtuelles Institut für Computerunterstützung in der klinischen Radiologie (2004–2006)

- DOT-MOBI

- HAMAM

Based on MeVisLab, the MedicalExplorationToolkit was developed to improve application development.[51] It is available as AddOn package for MeVisLab 1.5.2. and 1.6 on Windows.

MeVisLab can also be used to generate surface models of biomedical images and to export them in Universal 3D format for embedding in PDF files.[52]

27.8 Licensing

The MeVisLab SDK can be downloaded at no cost and without prior registration. The software can be used under three different license models:[53]

- MeVisLab SDK Unregistered: This license model applies if the MeVisLab SDK is used without an additional license file. Under this license, a restricted feature set is available. The terms of use are identical to those of the Non-commercial MeVisLab SDK (see below).

- Non-commercial MeVisLab SDK license: For strictly private use or for use at non-commercial institutions, such as universities, other academic institutions or non-profit organizations. Full feature set, requires a separate license file with costs.

- Commercial MeVisLab SDK license: For use at commercial companies, institutions or research laboratories. Full feature set, requires a separate license file with costs.

None of the above license models permits the redistribution of the MeVisLab SDK or parts thereof, or using MeVisLab or parts thereof as part of a commercial service or product.

The Fraunhofer MEVIS Release Modules are intellectual property of Fraunhofer MEVIS and strictly for non-commercial purposes.[53]

27.9 Related open source projects

27.9.1 MeVisLab public sources

As of MeVisLab 1.3, selected MeVisLab Standard modules are open source and available as MeVisLab Public Sources.[54] As of MeVisLab 2.0, these public sources are fully integrated in the MeVisLab SDK.

The source code is released under BSD license.

27.9.2 MeVisLab community and community sources

In the MeVisLab Community Project, open-source modules for MeVisLab are contributed by a number of institutions. Contributors as of 2010 are:

- Erasmus University Rotterdam, NL

- Medical Imaging Research Center, Katholieke Universiteit Leuven, BE

- Division of Image Processing (LKEB), Leiden University Medical Center, NL

- Computer Vision Laboratory, ETH Zurich, CH

- Institut für Simulation und Graphik, Universität Magdeburg, DE

- Center for Medical Image Science and Visualization (CMIV), University of Linköping, SE

- MeVis Medical Solutions AG

- Fraunhofer MEVIS

The source code is released under BSD or LGPL license and managed in a central repository on SourceForge. Continuous builds are offered for various platforms.

27.9.3   PythonQt

PythonQt is a Python script binding for the Qt framework. It was originally written to make MeVisLab scriptable and then published as open source in 2007 under LGPL. An introduction of PythonQt was published in Qt Quarterly, which also includes a comparison to Pyqt.

PythonQt sources and documentation are available from SourceForge.

27.10   Similar software projects

- Slicer (3DSlicer), an open source, multi-platform project for image analysis and scientific visualization; originally developed by the Surgical Planning Laboratory at the Brigham and Women's Hospital and the MIT Artificial Intelligence Laboratory

- SciRun, is an open source, multi-platform scientific problem solving environment (PSE) for modeling, simulation and visualization of scientific problems, developed at the Center for Integrative Biomedical Computing at the SCI, University of Utah

- XIP, the eXtensible Imaging Platform is an open source, multi-platform project for rapidly developing medical imaging applications from an extensible set of modular elements; originally developed at Siemens Corporate Research in Princeton

- MITK, the Medical Imaging Interaction Toolkit is an open source project for developing interactive medical image processing software, developed at the Deutsche Krebsforschungszentrum, Heidelberg

- Voreen, an open source, multi-platform volume rendering engine, maintained by the Visualization and Computer Graphics Research Group (VisCG) at the University of Muenster

- DeVIDE, an open source, multi-platform software for rapid prototyping, testing and deployment of visualisation and image processing algorithms, developed by the Visualisation group at the TU Delft.

- Amira, a commercial multi-platform software for visualization, analysis and manipulation of bio-medical data

27.11   See also

- Scientific visualization

- Graphical programming

- Medical imaging

27.12 References

[1] "MeVisLab History". Mevislab.de. Retrieved January 21, 2012.

[2] "MeVisLab 1.0 Release News". Mevislab.de. Retrieved January 21, 2012.

[3] "MeVisLab Features". Mevislab.de. Retrieved January 21, 2012.

[4] "MeVisLab Documentation". Mevislab.de. Retrieved January 21, 2012.

[5] "Ritter F, Boskamp T, Homeyer A, Laue H, Schwier M, Link F, Peitgen H-O. Medical Image Analysis: A Visual Approach. IEEE Pulse. 2011; 2(6):60–70". Ieeexplore.ieee.org. December 1, 2011. doi:10.1109/MPUL.2011.942929. Retrieved January 21, 2012.

[6] Link F, König M, Peitgen H-O; Multi-Resolution Volume Rendering with per Object Shading. In: Kobbelt L, Kuhlen T, Westermann R, eds. Vision Modelling and Visualization. Berlin, Aachen: Aka; 2006:185–191

[7] SoGVR Renderer Module Documentation

[8] "Heckel F, Schwier M, Peitgen H-O; Object-oriented application development with MeVisLab and Python; Lecture Notes in Informatics (Informatik 2009: Im Focus das Leben), 2009, 154, pp. 1338–1351" (PDF). Retrieved January 21, 2012.

[9] "Open Inventor Reference". Mevislab.de. Retrieved January 21, 2012.

[10] Rexilius J, Jomier J, Spindler W, Link F, König M, Peitgen H-O; Combining a Visual Programming and Rapid Prototyping Platform with ITK. In: Bildverarbeitung für die Medizin. Berlin: Springer, 2005: 460–464

[11] "Rexilius J, Spindler W, Jomier J, Koenig M, Hahn H-K, Link F, Peitgen H-O; A Framework for Algorithm Evaluation and Clinical Application Prototyping using ITK. The Insight Journal 2005; ISC/NA-MIC/MICCAI Workshop on Open-Source Software". Insight-journal.org. Retrieved January 21, 2012.

[12] "Bitter I, van Uitert R, Wolf I, Ibanez L, Kuhnigk J-M; Comparison of Four Freely Available Frameworks for Image Processing and Visualization That Use ITK; IEEE Trans Visual Comput Graphics,13(3): 483–493, 2007 May/June". Ieeexplore.ieee.org. March 19, 2007. doi:10.1109/TVCG.2007.1001. Retrieved January 21, 2012.

[13] Koenig M, Spindler W, Rexilius J, Jomier J, Link F, Peitgen H-O; Embedding VTK and ITK into a Visual Programming and Rapid Prototyping Platform. In: Proceedings of SPIE – Volume 6141 Medical Imaging 2006 Image Processing. Bellingham: SPIE, 2006: in press

[14] VTK Module Reference

[15] "MeVisLab Reference Manual". Mevislab.de. September 3, 2011. Retrieved January 21, 2012.

[16] Release Notes MeVisLab 1.0

[17] Release Notes MeVisLab 1.1

[18] Release Notes MeVisLab 1.2

[19] Release Notes MeVisLab 1.3

[20] "Release Notes MeVisLab Public Sources". Mevislab.de. Retrieved January 21, 2012.

[21] "Release Notes MeVisLab ITK/VTK Integration". Mevislab.de. Retrieved January 21, 2012.

[22] Release Notes MeVisLab 1.4

[23] Release Notes MeVisLab 1.5

[24] Release Notes MeVisLab 1.6

[25] Release Notes MeVisLab 2.0

[26] Release Notes MeVisLab 2.1

[27] "Release Notes MeVisLab 2.2". Mevislab.de. Retrieved January 21, 2012.

[28] "Release Notes MeVisLab 2.3". Mevislab.de. Retrieved July 26, 2012.

[29] "Release Notes MeVisLab 2.4". Mevislab.de. Retrieved February 12, 2013.

[30] "Release Notes MeVisLab 2.5". Mevislab.de. Retrieved October 17, 2013.

[31] "Release Notes MeVisLab 2.6". Mevislab.de. Retrieved June 1, 2014.

[32] http://isgwww.cs.uni-magdeburg.de/visualisierung/wiki/lib/exe/fetch.php?media=files:animation_exploration:muehler_2010_eurovis.pdf

[33] "Rieder C, Schwier M, Weihusen A, Zidowitz S, Peitgen, H-O; Visualization of Risk Structures for Interactive Planning of Image Guided Radiofrequency Ablation of Liver Tumors; SPIE Medical Imaging: Visualization, Image-Guided Procedures, and Modeling, Orlando, 2009" (PDF). Retrieved January 21, 2012.

[34] "Zidowitz S, Hansen C, Schlichting S, Kleemann M, Peitgen, H-O; Software assistance for intra-operative guidance in liver surgery; World Congress on Medical Physics and Biomedical Engineering 2009. Vol.6: Surgery, minimal invasive interventions, edoscopy and image guided therapy, pages 205–208, 2009". Springerlink.com. Retrieved January 21, 2012.

[35] "Hansen C, Lindow B, Zidowitz S, Schenk A, Peitgen H-O; Towards Automatic Generation of Resection Surfaces for Liver Surgery Planning; Proceedings of Computer Assisted Radiology and Surgery (CARS) 2010, 5 (Suppl. 1), pp. 119–120" (PDF). Retrieved January 21, 2012.

[36] "Liver projects at Fraunhofer MEVIS". Mevis.de. Retrieved January 21, 2012.

[37] "Dicken V, Kuhnigk J-M, Bornemann L, Zidowitz S, Krass S, Peitgen H-O; Novel CT data analysis and visualization techniques for risk assessment and planning of thoracic surgery in oncology patients; in H.U. Lemke, K. Inamura, K. Doi, M.W. Vannier, and A.G. Farman, editors, Proc CARS: Computer Assisted Radiology and Surgery, volume 1281 of Computer Assisted Radiology and Surgery, pages 783–787, Amsterdam, 2005". Dx.doi.org. June 22, 2005. Retrieved January 21, 2012.

[38] "Lung projects at Fraunhofer MEVIS". Mevis.de. Retrieved January 21, 2012.

[39] "Rieder C, Görge H-H, Ritter F, Hahn H-K, Peitgen H-O; Efficient Visualization of Risk Structures along Virtual Access Paths for Neurosurgical Planning; 59th Annual Meeting of the German Society of Neurosurgery (DGNC), Würzburg, 2008" (PDF). Retrieved January 21, 2012.

[40] "Neuro projects at Fraunhofer MEVIS". Mevis.de. Retrieved January 21, 2012.

[41] "Breast projects at Fraunhofer MEVIS". Mevis.de. Retrieved January 21, 2012.

[42] Hahn H-K, Harz M-T, Seyffarth H, Zöhrer F, Böhler T, Filippatos K, Wang L, Homeyer A, Ritter F, Laue H, Günther M, Twellmann T, Tabár L, Bick U, Schilling K; Concepts for Efficient and Reliable Multi-Modal Breast Image Reading; International Workshop on Digital Mammography (IWDM 2010, June 16–18, Girona, Spain), pp.

[43] "Visual computing for medical diagnosis and treatment" (PDF). Retrieved January 21, 2012.

[44] "Standardized evaluation methodology and reference database for evaluating coronary artery centerline extraction algorithms" (PDF). Retrieved January 21, 2012.

[45] "Cardio projects at Fraunhofer MEVIS". Mevis.de. Retrieved January 21, 2012.

[46] "Bolte H, Jahnke T, Schafer F-K, Wenke R, Hoffmann B, Freitag-Wolf S, Dicken V, Kuhnigk J-M, Lohmann J, Voss S, Knoss N, Heller M, Biederer J; Interobserver-variability of lung nodule volumetry considering different segmentation algorithms and observer training levels; Eur J Radiol, 64(2): 285–295, 2007 April". Ncbi.nlm.nih.gov. October 3, 2011. Retrieved January 21, 2012.

[47] "Rieder C, Weihusen A, Schumann C, Zidowitz S, Peitgen H-O; Visual Support for Interactive Post-Interventional Assessment of Radiofrequency Ablation Therapy; Computer Graphics Forum (Special Issue on Eurographics Symposium on Visualization) 29, 3 (1093–1102), 2010" (PDF). Retrieved January 21, 2012.

[48] "Klein J, Bartz D, Friman O, Hadwiger M, Preim B, Ritter F, Vilanova A, Zachmann G; Advanced Algorithms in Medical Computer Graphics; Eurographics 2008, Crete, April 14–18. State-of-the-Art Report (EG-STAR'08)" (PDF). Retrieved January 21, 2012.

[49] Felix Ritter. "Ritter F; Visual Programming for Prototyping of Medical Applications; IEEE Visualization 2007, Sacramento, October 28 – November 1. Tutorial: "Introduction to Visual Medicine: Techniques, Applications and Software" by Dirk Bartz, Klaus Mueller, Felix Ritter, Bernhard Preim, and Karel Zuiderveld". Mevis-research.de. Retrieved January 21, 2012.

[50] Bornemann L, Dicken V, Kuhnigk J-M, Beyer F, Shin H, Bauknecht C, Diehl V, Fabel-Schulte M, Meier S, Kress O, Krass S, Peitgen H-O; Software Assistance for Quantitative Therapy Monitoring in Oncology; Proc Workshop on Medical Image Processing: Challenges in Clinical Oncology: 40–46, 2006]

[51] "Mühler K, Tietjen C, Ritter F, Preim B; The Medical Exploration Toolkit: An Efficient Support for Visual Computing in Surgical Planning and Training; IEEE Transactions on Visualization and Computer Graphics (133–146), Los Alamitos, CA, USA, 2010" (PDF). Retrieved January 21, 2012.

[52] "Simplified Generation of Biomedical 3D Surface Model Data for Embedding into 3D Portable Document Format (PDF) Files for Publication and Education". Retrieved 2014-02-14.

[53] "MeVisLab Versions and Licensing". Mevislab.de. Retrieved January 21, 2012.

[54] "MeVisLab Public Sources". Mevislab.de. Retrieved January 21, 2012.

27.13   Further reading

- MeVisLab Publications

- Medical Image Analysis: A Visual Approach

- Using VTK in MeVisLab (PDF)

- Object-oriented application development with MeVisLab and Python

- Entwicklung eines Werkzeugs zur Koordinaten- und Grauwertinterpolation von MRI- und PET-Daten in der objektorientierten Umgebung MeVisLab (Diplomarbeit)

27.14   External links

- MeVisLab Home Page

- MeVisLab Community Sources

- MeVisLab Support Forum

- MeVis Medical Solutions AG

- Fraunhofer MEVIS

- MeVisLab for Mac OS X, Intro and Demos

Chapter 28

Mocolo

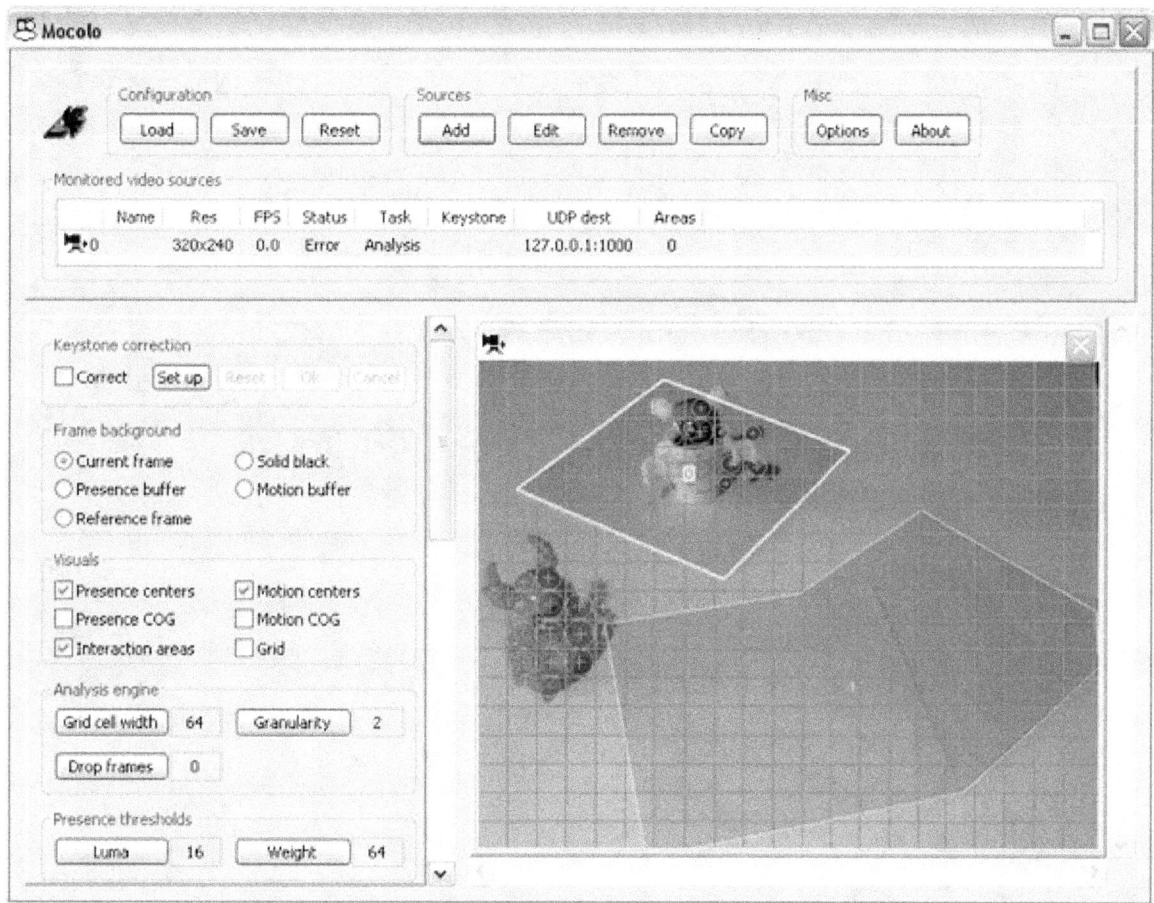

Mocolo screenshot

Mocolo is a proprietary video analysis server for Microsoft Windows, mostly used in interactive environments. It monitors video streams, finds interactions (presences or movements) and sends analysis results to a list of client applications in form of Open Sound Control (OSC) packets. Key features: lightweight performance, capability to analyze several video sources at the same time, connectivity toward any OSC capable application. Mocolo makes use of small amounts of CPU time and no use of the GPU, leaving enough resources for other client applications running on the same machine.

28.1 External links

- Mocolo homepage

- Xtend3dLab projects page

Chapter 29

OpenCV

OpenCV (*Open Source Computer Vision*) is a library of programming functions mainly aimed at real-time computer vision, originally developed by Intel research center in Nizhny Novgorod (Russia), later supported by Willow Garage and now maintained by Itseez.[1] The library is cross-platform and free for use under the open-source BSD license.

29.1 History

Officially launched in 1999, the OpenCV project was initially an Intel Research initiative to advance CPU-intensive applications, part of a series of projects including real-time ray tracing and 3D display walls. The main contributors to the project included a number of optimization experts in Intel Russia, as well as Intel's Performance Library Team. In the early days of OpenCV, the goals of the project were described as:

- Advance vision research by providing not only open but also optimized code for basic vision infrastructure. No more reinventing the wheel.

- Disseminate vision knowledge by providing a common infrastructure that developers could build on, so that code would be more readily readable and transferable.

- Advance vision-based commercial applications by making portable, performance-optimized code available for free—with a license that did not require to be open or free themselves.

The first alpha version of OpenCV was released to the public at the IEEE Conference on Computer Vision and Pattern Recognition in 2000, and five betas were released between 2001 and 2005. The first 1.0 version was released in 2006. In mid-2008, OpenCV obtained corporate support from Willow Garage, and is now again under active development. A version 1.1 "pre-release" was released in October 2008.

The second major release of the OpenCV was on October 2009. OpenCV 2 includes major changes to the C++ interface, aiming at easier, more type-safe patterns, new functions, and better implementations for existing ones in terms of performance (especially on multi-core systems). Official releases now occur every six months[2] and development is now done by an independent Russian team supported by commercial corporations.

In August 2012, support for OpenCV was taken over by a non-profit foundation OpenCV.org, which maintains a developer[3] and user site.[4]

29.2 Applications

OpenCV's application areas include:

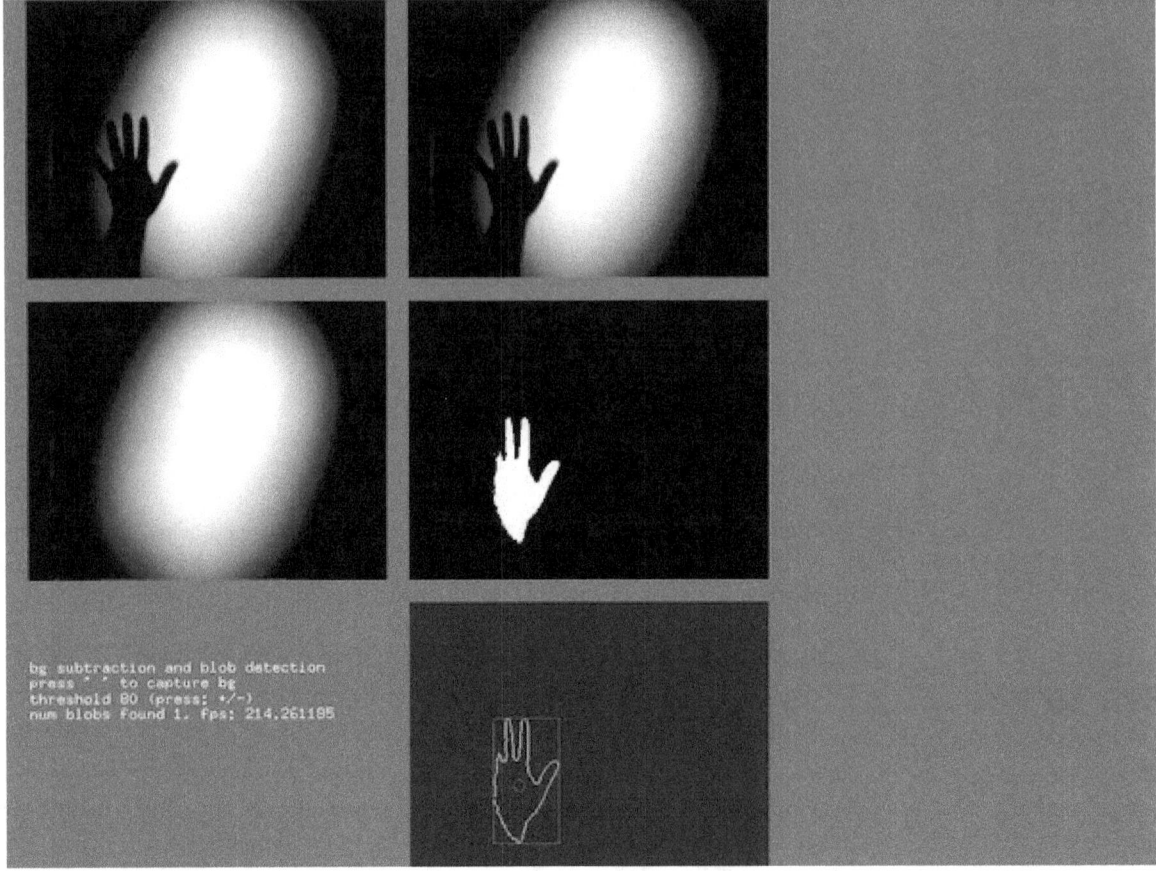

openFrameworks running the OpenCV add-on example

- 2D and 3D feature toolkits

- Egomotion estimation

- Facial recognition system

- Gesture recognition

- Human–computer interaction (HCI)

- Mobile robotics

- Motion understanding

- Object identification

- Segmentation and recognition

- Stereopsis stereo vision: depth perception from 2 cameras

- Structure from motion (SFM)

- Motion tracking

- Augmented reality

To support some of the above areas, OpenCV includes a statistical machine learning library that contains:

- Boosting (meta-algorithm)

- Decision tree learning

- Gradient boosting trees

- Expectation-maximization algorithm

- k-nearest neighbor algorithm

- Naive Bayes classifier

- Artificial neural networks

- Random forest

- Support vector machine (SVM)

29.3   Programming language

OpenCV is written in C++ and its primary interface is in C++, but it still retains a less comprehensive though extensive older C interface. There are bindings in Python, Java and MATLAB/OCTAVE. The API for these interfaces can be found in the online documentation.[5] Wrappers in other languages such as C#, Perl,[6] Ch,[7] and Ruby have been developed to encourage adoption by a wider audience.

All of the new developments and algorithms in OpenCV are now developed in the C++ interface.

29.4   Hardware Acceleration

If the library finds Intel's Integrated Performance Primitives on the system, it will use these proprietary optimized routines to accelerate itself.

A CUDA-based GPU interface has been in progress since September 2010.[8]

An OpenCL-based GPU interface has been in progress since October 2012,[9] documentation for version 2.4.9.0 can be found at docs.opencv.org.[10]

29.5   OS support

OpenCV runs on a variety of platforms. Desktop: Windows, Linux, OS X, FreeBSD, NetBSD, OpenBSD; Mobile: Android, iOS, Maemo,[11] BlackBerry 10.[12] The user can get official releases from SourceForge or take the latest sources from GitHub.[13] OpenCV uses CMake.

29.6   See also

- AForge.NET, a computer vision library for the Common Language Runtime (.NET Framework and Mono).

- ROS (Robot Operating System). OpenCV is used as the primary vision package in ROS.

- VXL, an alternative library written in C++.

- Integrating Vision Toolkit (IVT), a fast and easy-to-use C++ library with an optional interface to OpenCV.

- CVIPtools, a complete GUI-based computer-vision and image-processing software environment, with C function libraries, a COM-based DLL, along with two utility programs for algorithm development and batch processing.

- OpenNN, an open-source neural networks library written in C++.

29.7 References

[1] Itseez leads the development of the renowned computer vision library OpenCV. http://itseez.com

[2] OpenCV change logs: http://code.opencv.org/projects/opencv/wiki/ChangeLog

[3] OpenCV Developer Site: http://code.opencv.org

[4] OpenCV User Site: http://opencv.org/

[5] OpenCV C interface: http://docs.opencv.org

[6] CPAN: http://search.cpan.org/~{}yuta/Cv-0.29/

[7] Ch OpenCV: http://www.softintegration.com/products/thirdparty/opencv/

[8] Cuda GPU port: http://opencv.org/platforms/cuda.html

[9] OpenCL Announcement: http://opencv.org/opencv-v2-4-3rc-is-under-way.html

[10] OpenCL-accelerated Computer Vision API Reference: http://docs.opencv.org/modules/ocl/doc/ocl.html

[11] Maemo port: https://garage.maemo.org/projects/opencv

[12] BlackBerry 10 (partial port): https://github.com/blackberry/OpenCV

[13] https://github.com/Itseez/opencv

29.8 External links

- OpenCV on SourceForge.net
- Documentation of OpenCV
- Introduction to programming with OpenCV
- A list of other open source computer vision codes and libraries
- Chroma Key Background Subtraction - OpenCV

Chapter 30

Pfinder

Pfinder is a computer vision system which detects features in video images in order to recognize human figures and their movements and gestures. Pfinder was designed by Wren, et al.[1] of the MIT Media Laboratory in 1997. As described by its authors, **Pfinder** is a "real-time system for tracking people and interpreting their behavior". The system improves upon previous works by not only identifying the boundaries of a person in the image, but also analyzing the regions inside the boundaries and relating them to the known structure of the human body. As an example, Pfinder can track a person's head and hands, and can determine the pose of the body and recognize gestures.

30.1 Limitations

- Pfinder does not cope with multi-modal backgrounds, in which a histogram of the pixel intensity contains more than one distinct peak.[2]

- While it can handle small or gradual changes in lighting, it does not react well to large, sudden lighting changes. When there are large lighting changes, the system mistakenly labels them as a part of the foreground, and therefore tries to incorporate them into the human figure model.[1]

- Pfinder is not able to handle multiple people in the same image well. While the blobs representing each person would be detected, the system would attempt to analyze them as one distinct human figure.[1]

30.2 Applications

- Pfinder has been applied to the recognition of American Sign Language[1]

30.3 See also

- Background subtraction

30.4 References

[1] Wren, Christopher Richard; Ali Azarbayejani; Ali Azarbayejani; Alex Paul Pentland (July 1997). "Pfinder: Real-Time Tracking of the Human Body" (PDF). *IEEE TRANSACTIONS ON PATTERN ANALYSIS AND MACHINE INTELLIGENCE* **19** (7): 780–785. Retrieved 24 July 2012.

[2] Picardi, Massimo (15 April 2004). "Background subtraction techniques: a review" (PDF). University of Technology, Sydney. p. 13. Retrieved 24 July 2012. It does not cope with multimodal backgrounds

Chapter 31

Pipeline Pilot

Pipeline Pilot is the authoring tool for the Accelrys Enterprise Platform. It is a scientific visual and dataflow programming language, used in various scientific domains, such as cheminformatics and QSAR,[1][2][3] Next Generation Sequencing,[4] image analysis,[5][6] text analytics.[7]

31.1 History

Originally created in 1999 by SciTegic, Pipeline Pilot is now developed by Accelrys.

Pipeline Pilot was used at first in the pharmaceutical and biotechnology industries and by academics and government agencies. Then other industries started to adopt it, but always in science driven sectors such as Chemicals, Energy, Consumer Packaged Goods, Aerospace, Automotive, Electronics.

31.2 Basic introduction

Pipeline Pilot includes contextual help that is searchable and interactive; users should refer to it. Reviewing the examples and the documentation is the best place to start.

31.2.1 Components, pipelines, protocols and data records

The graphical user interface, called the Pipeline Pilot Professional Client, allows users to drag and drop components, connect them together in pipelines, and save the application developed as a protocol.

Think of the components as nodes of a directed graph: each one has a specific task on the data. Users have the choice to use predefine components, or to develop their own: components can be chosen from the library, configured, redesigned, or even created from scratch and documented at will. When a new component is made by collapsing a few components together, we also call it a subprotocol.

In a typical protocol, the reading components (on the left) send the data records through the pipelines (to the right) for further process, analysis, and reporting.

31.2.2 Component Collections

The components are organised by science in collections.[8]

The most interesting protocols are often those mixing collections:

114

- Determine the IC50 by calculating the dose–response relationship directly from information extracted from high-content screening assay images, associated with dilution in the plate layout and chemistry information about the tested compounds (Imaging, Chemistry, Plate Data Analytics)

- Suggest which scientific article to read next, based on a Bayesian model built upon text fingerprints and user's reading list or papers ranking (Text Analytics, Data Modeling)

- Access experiment methods and results from electronic laboratory notebook or laboratory information management system, and report for resource capacity planning (Integration, Reporting)

31.2.3 PilotScript

Many custom script components are available in Pipeline Pilot, allowing experts to include their code directly into the pipelines and maintain a library of components based on their preferred language, such as Perl, Java, VBScript, .NET, JavaScript, Python, Matlab, etc.[9]

Another option is to use PilotScript, the internal scripting language of Pipeline Pilot, which syntax is based on PLSQL. It can be used in components such as the *Custom Manipulator (PilotScript)* or the *Custom Filter (PilotScript)*.

Hello := "Hello World!";

This script above will add, to each data record passing through its component, a property named Hello containing the string "Hello World!".

31.3 Community

The community forum allows users to share ideas, components, and protocols.

Companies such as ACD/Labs,[10] BioSolveIT,[11] ChemAxon,[12] Cosmologic,[13] Cresset,[14] Linguamatics,[15] Molecular Discovery,[16] and Molecular Networks,[17] partner with Accelrys[18] to develop and provide component collections to interface with their technologies, allowing automation from within Pipeline Pilot.

Oxford Nanopore Technologies offers Pipeline Pilot NGS collection as the preferred and supported solution for secondary and higher level data analysis in their GridION system.[19][20]

31.4 References

[1] Hassan, Moises; Brown, Robert D.; Varma-O'Brien, Shikha; Rogers, David (2007). "Cheminformatics Analysis and Learning in a Data Pipelining Environment". *ChemInform* **38** (12). doi:10.1002/chin.200712278. ISSN 0931-7597.

[2] Hu, Ye; Lounkine, Eugen; Bajorath, Jürgen (2009). "Improving the Search Performance of Extended Connectivity Fingerprints through Activity-Oriented Feature Filtering and Application of a Bit-Density-Dependent Similarity Function". *ChemMedChem* **4** (4): 540–548. doi:10.1002/cmdc.200800408. ISSN 1860-7179.

[3] Warr, Wendy A. (2012). "Scientific workflow systems: Pipeline Pilot and KNIME". *Journal of Computer-Aided Molecular Design* **26** (7): 801–804. doi:10.1007/s10822-012-9577-7. ISSN 0920-654X. PMC 3414708. PMID 22644661.

[4] "Accelrys Enters Next Generation Sequencing Market with NGS Collection for Pipeline Pilot". Business Wire. Retrieved 15 February 2013.

[5] Rabal, Obdulia; Link, Wolfgang; G. Serelde, Beatriz; Bischoff, James R.; Oyarzabal, Julen (2010). "An integrated one-step system to extract, analyze and annotate all relevant information from image-based cell screening of chemical libraries". *Molecular BioSystems* **6** (4): 711–20. doi:10.1039/b919830j. ISSN 1742-206X. PMID 20237649.

[6] Paveley, Ross A.; Mansour, Nuha R.; Hallyburton, Irene; Bleicher, Leo S.; Benn, Alex E.; Mikic, Ivana; Guidi, Alessandra; Gilbert, Ian H.; Hopkins, Andrew L.; Bickle, Quentin D. (2012). "Whole Organism High-Content Screening by Label-Free, Image-Based Bayesian Classification for Parasitic Diseases". *PLoS Neglected Tropical Diseases* **6** (7): e1762. doi:10.1371/journal.pntd.0001762. ISSN 1935-2735.

[7] Vellay, SG; Latimer, NE; Paillard, G (2009). "Interactive text mining with Pipeline Pilot: a bibliographic web-based tool for PubMed". *Infectious disorders drug targets* **9** (3): 366–74. doi:10.2174/1871526510909030366. PMID 19519489.

[8] "Pipeline Pilot Component Collections". Accelrys. Retrieved 26 January 2013.

[9] "Pipeline Pilot Integration Component Collection Datasheet" (PDF). Accelrys. Retrieved 8 February 2013.

[10] "ACD/Labs Component Collection for Accelrys Pipeline Pilot". ACD/Labs. Retrieved 26 January 2013.

[11] "Pipeline Pilot Interfaces". BioSolveIT. Retrieved 26 January 2013.

[12] "Pipeline Pilot Components". ChemAxon. Retrieved 26 January 2013.

[13] "Integration of COSMOfrag and COSMOsim in Pipeline Pilot". Cosmologic. Retrieved 26 January 2013.

[14] "Pipeline Pilot Protocols". Cresset. Retrieved 12 February 2013.

[15] "I2E Pipeline Pilot Integration". Linguamatics. Retrieved 26 January 2013.

[16] Cross, Simon; Sforna, Gianluca; Vianello, Riccardo. "Technical Note: Integration of MoKa, VolSurf+, and MetaSite into Pipeline Pilot" (PDF). Molecular Discovery. Retrieved 26 January 2013.

[17] "Components for Pipeline Pilot". Molecular Networks. Retrieved 26 January 2013.

[18] "Accelrys Partners". Accelrys. Retrieved 26 January 2013.

[19] "Oxford Nanopore selects Accelrys Pipeline Pilot NGS Collection as preferred analysis platform". News-Medical.net. 16 March 2011. Retrieved 15 February 2013.

[20] "DNA sequencing informatics, GridION™ Nodes". Oxford Nanopore Technologies. Retrieved 26 January 2013.

Chapter 32

RapidMiner

RapidMiner is a software platform developed by the company of the same name that provides an integrated environment for machine learning, data mining, text mining, predictive analytics and business analytics. It is used for business and industrial applications as well as for research, education, training, rapid prototyping, and application development and supports all steps of the data mining process including results visualization, validation and optimization.[1] RapidMiner is developed on a business source model which means only the previous version of the software is available under an OSI-certified open source license on Sourceforge.[2] A Starter Edition is available for free download, a Personal Edition is offered for US$999, a Professional Edition is $2,999 and pricing for the Enterprise Edition is available from the developer.[3]

32.1 History

RapidMiner, formerly known as YALE (Yet Another Learning Environment), was developed starting in 2001 by Ralf Klinkenberg, Ingo Mierswa, and Simon Fischer at the Artificial Intelligence Unit of the Technical University of Dortmund.[4] Starting in 2006, its development was driven by Rapid-I, a company founded by Ingo Mierswa and Ralf Klinkenberg in the same year.[5] In 2007, the name of the software was changed from YALE to RapidMiner and the company Rapid-I GmbH was incorporated.[6]

32.2 Description

RapidMiner uses a client/server model with the server offered as Software as a Service or on cloud infrastructures.[7]

According to Bloor Research, RapidMiner provides 99% of an advanced analytical solution through template-based frameworks that speed delivery and reduce errors by nearly eliminating the need to write code. RapidMiner provides data mining and machine learning procedures including: data loading and transformation (Extract, transform, load (ETL)), data preprocessing and visualization, predictive analytics and statistical modeling, evaluation, and deployment. RapidMiner is written in the Java programming language. RapidMiner provides a GUI to design and execute analytical workflows. Those workflows are called "Process" in RapidMiner and they consist of multiple "Operators". Each operator is performing a single task within the process and the output of each operator forms the input of the next one. Alternatively, the engine can be called from other programs or used as an API. Individual functions can be called from the command line. RapidMiner provides learning schemes and models and algorithms from Weka and R scripts that can be used through extensions.[8]

RapidMiner functionality can be extended with additional plugins which are made available via RapidMiner Marketplace. The Rapid Miner Extensions marketplace provides a platform for developers to create data analysis algorithms and publish them to the community.[9] Previous versions RapidMiner are distributed under the AGPL open source license. As of 15 May 2015, Version 5.3.013 is available as Open Source, while RapidMiner 6 was released in 2013. The development model of RapidMiner is not open, the company only infrequently gives out the source code of a previous version.

With version 6.0, RapidMiner started to offer new application wizards addressed to business analysts needs for predictive analytics.[10]

32.3   Adoption

In 2014, Gartner Research placed RapidMiner in the leader quadrant of its Magic Quadrant for Advanced Analytics. The report described RapidMiner's strengths as a "platform that supports an extensive breadth and depth of functionality, and with that it comes quite close to the market Leaders."[11] In the 2014 and 2013 annual software poll KDnuggets ranked RapidMiner the most popular data analytics software with the poll's respondents citing the software package as the tool they use.[12][13] RapidMiner received one of the strongest satisfaction ratings in the 2011 Rexer Analytics Data Miner Survey.[14] RapidMiner has received over 3 million total downloads and has over 200,000 users including eBay, Intel, PepsiCo and Kraft Foods as paying customers. RapidMiner claims to be the market leader in the software for predictive data analytics services against competitors such as Revolution Analytics, SAS, Predixion Software, SQL Server, StatSoft and IBM.[15]

32.4   Developer

About 50 developers worldwide participate in the development of the open source RapidMiner with the majority of the contributors being employees of RapidMiner.[16] The company that develops RapidMiner software recently changed its name from Rapid-I to RapidMiner and received a $5 million series A funding with participation from European venture capital firms Earlybird Venture Capital and Open Ocean Capital. The company stated that the funding will be used to build out the development and marketing teams.[17] Open Ocean partner Michael "Monty" Widenius is a founder of MySQL.

32.5   References

[1] Markus Hofmann, Ralf Klinkenberg, "RapidMiner: Data Mining Use Cases and Business Analytics Applications (Chapman & Hall/CRC Data Mining and Knowledge Discovery Series)," *CRC Press*, October 25, 2013.

[2] "The core of RapidMiner is open source". RapidMiner. Retrieved 18 July 2014.

[3] RapidMiner 6 Review, Butler Analytics, November 22, 2013.

[4] Guido Deutsch, "RapidMiner from Rapid-I at CeBIT 2010," *Data Mining Blog*, March 18, 2010.

[5] "Interview with RapidMiner's Ingo Mierswa, Ralf Klinkenberg", *KDnuggets*, February, 2010.

[6] "Free Data Mining Software: RapidMiner 4.0 (formerly YALE)", *KDNuggets*, August 7, 2007.

[7] David Norris, "RapidMiner - a potential game changer," *IT-Director.com*, November 22, 2013.

[8] David Norris, "RapidMiner - a potential game changer," Bloor Research, November 13, 2013.

[9] Ajay Ohri, "Interview with Rapid-I Ingo Mierswa and Simon Fischer," *KDnuggets*, August 2011.

[10] RapidMiner 6 Review, Butler Analytics, November 22, 2013.

[11] "RapidMiner: Leader in Gartner Research Magic Quadrant for Advanced Analytics Platforms," *Garnter*, February 24, 2014.

[12] "KDnuggets Annual Software Poll:RapidMiner and R vie for first place," *KDnuggets*, June 2013.

[13] "KDnuggets 15th Annual Software Poll:RapidMiner continues to lead.," *KDnuggets*, June 2014.

[14] "2011 Data Miner Survey," Rexer Analytics.

[15] Ingrid Lunden, "German Predictive Analytics Startup Rapid-I Rebrands As RapidMiner, Takes $5M From Open Ocean, Early-bird To Tackle The U.S. Market," *TechCrunch*, November 4, 2013.

[16] Evan Quinn, "Is Rapid-I the Hidden Giant of Analytics?," QuinnSight Research, June 17, 2013.

[17] Andrew Brust, "Rapid-I gets funded, re-brands as RapidMiner," *ZDNet*, November 4, 2013.

32.6 External links

- Official website

- RapidMiner on SourceForge.net

Chapter 33

RoboRealm

RoboRealm is an application for use in computer vision, image analysis, and robotic vision systems. RoboRealm provides a Windows based GUI for experimenting with different modules that can be assembled in custom ways to achieve a desired result. The main goal behind RoboRealm is to translate visual input into actuator commands that can be used to move robots or trigger actions based on what a machine sees.

Many different algorithms are implemented: Blob and Particle Algorithms; Color Algorithms; Edge Detection; Filters; among other common computer vision functionality. As a tool for computer vision applied to robotics, it also includes methods for robot localization and basic navigation skills, and integration with common robotic sensors. There are also third party modules that like the AVM Navigator module that provides autonomous robot navigation based on the visual landmarks.

RoboRealm's GUI interface allows for intuitive exploration of very advanced concepts. Machine vision is a very complex field that requires many complex mathematical and programming concepts in order to be successful. RoboRealm provides the ability to 'play' with these algorithms in order to develop a better intuitive model on what an algorithm can do and how it can be successfully applied to a project. Values that change the behavior of vision algorithms can be adjusted with immediate changes to how the resulting image is processed. This streamlines the tweak, compile, run, rethink process needed when researching vision based algorithms using source code. RoboRealm connects to many input devices including webcams, static images, video files, IP cameras and various other imaging devices. Because of this it is possible to use image content from various sources for test purposes.

Vision systems are typically part of a larger robotic solution. Due to this RoboRealm includes various way in which it can be extended in order to accomplish a custom task. These extensions include embedded modules that allow scripting in various languages (VBScript, Python, and CScript), a plugin architecture that allows individuals to create custom modules that are incorporated into RoboRealm processing pipeline (DLL, pipes) and an API that provides server based functionality through simple XML remote procedure calls. Augmenting these three extension abilities are several modules that can produce and consume information generated by RoboRealm such that this information can be incorporated into your own system. "Read Variables", "Write Images", "Clipboard", etc. are a few that further enhance information exchange within RoboRealm. For further information about integration see Integration with RoboRealm

RobotRealm requires the Microsoft Windows operating system but leverages this by supporting a large variety of robotic and programming platforms, such as Surveyor, Lego Mindstorms NXT, Vex Robotics Design System, iRobot Create and Microsoft Robotics Developer Studio.

33.1 Highlights

- Easy to Use GUI Interface
- Hundreds of Image Processing Modules
- Realtime Parameter Changes

- Inexpensive Vision Application

- Fully Supported Server API

- Multiple Image Sources (IEEE 1394 firewire, webcam, movie files, web images, etc.)

- Multiple Output Interfaces (Disk, Web, FTP, Email, etc.)

- Plugin Framework for Custom Modules

33.2 Usage

RoboRealm can be accessed using the GUI or the API. The GUI provides intuitive access to complex algorithms. Once understood, the API, which is based on a socket based communication protocol, allows other systems to access RoboRealm from remote locations and using different languages to incorporate its capabilities into custom implementations.

The protocol is based on a tagged language.

<request> <get_dimension>IMAGE_NAME</get_dimension> </request> <response> <width>IMAGE_WIDTH</width> <height>IMAGE_HEIGHT</height> </response>

33.3 External links

- Official website

- Development of the forward kinematics for robot fingers by using RoboRealm

- Using the BOL-BOT with RoboRealm Vision System

- Tracking cars on a freeway using Roborealm & a webcam

- How to make Barcode Reading using Roborealm

- Roborealm sidewalk following

- Hexapod + Roborealm

- Humanoid walking and tracking a ball: Lego Mindstorms NXT

- Simple object tracking in Roborealm - Tutorial

- Highest point data in Roborealm - Tutorial

Chapter 34

Robot Operating System

Robot Operating System (**ROS**) is a collection of software frameworks for robot software development, (see also Robotics middleware) providing operating system-like functionality on a heterogeneous computer cluster. ROS provides standard operating system services such as hardware abstraction, low-level device control, implementation of commonly used functionality, message-passing between processes, and package management. Running sets of ROS-based processes are represented in a graph architecture where processing takes place in nodes that may receive, post and multiplex sensor, control, state, planning, actuator and other messages. Despite the importance of reactivity and low latency in robot control, ROS, itself, is *not* a Realtime OS, though it is possible to integrate ROS with realtime code.[2]

Software in the ROS Ecosystem can be separated into three groups:

- language- and platform-independent tools used for building and distributing ROS-based software;

- ROS client library implementations such as roscpp, rospy, and roslisp;

- packages containing application-related code which uses one or more ROS client libraries.

Both the language-independent tools and the main client libraries (C++, Python, LISP) are released under the terms of the BSD license, and as such are open source software and free for both commercial and research use. The majority of other packages are licensed under a variety of open source licenses. These other packages implement commonly used functionality and applications such as hardware drivers, robot models, datatypes, planning, perception, simultaneous localization and mapping, simulation tools, and other algorithms.

The main ROS client libraries (C++, Python, LISP) are geared toward a Unix-like system, due primarily because of their dependence on large collections of open-source software dependencies. For these client libraries, Ubuntu Linux is listed as "Supported" while other variants such as Fedora Linux, Mac OS X, and Microsoft Windows are designated "Experimental" and are supported by the community.[3] The native Java ROS client library, rosjava, however, does not share these limitations and has enabled ROS-based software to be written for the Android OS.[4] rosjava has also enabled ROS to be integrated into an officially-supported MATLAB toolbox which can be used on Linux, Mac OS X, and Microsoft Windows.[5] A JavaScript client library, roslibjs has also been developed which enables integration of software into a ROS system via any standards-compliant web browser.

34.1 History

ROS was originally developed in 2007 under the name *switchyard* by the Stanford Artificial Intelligence Laboratory in support of the Stanford AI Robot STAIR (STanford AI Robot) [6][7] project.

Description of STAIR on its homepage : —

> Our single robot platform will integrate methods drawn from all areas of AI, including machine learning, vision, navigation, planning, reasoning, and speech/natural language processing. This is in distinct contrast to the 30-year trend of working on fragmented AI sub-fields, and will be a vehicle for driving research towards true integrated AI.

From 2008 until 2013, development was performed primarily at Willow Garage, a robotics research institute/incubator. During that time, researchers at more than twenty institutions collaborated with Willow Garage engineers in a federated development model.[8][9]

In February 2013, ROS stewardship transitioned to the Open Source Robotics Foundation.[10] In August 2013, a blog posting[11] announced that Willow Garage would be absorbed by another company started by its founder, Suitable Technologies. The support responsibilities for the PR2 created by Willow Garage were also subsequently taken over by Clearpath Robotics.[12]

34.2 Applications

ROS areas include:

- A master coordination node
- Publishing or subscribing to data streams: images, stereo, laser, control, actuator, contact ...
- Multiplexing information
- Node creation and destruction
- Nodes are seamlessly distributed, allowing distributed operation over multi-core, multi-processor, GPUs and clusters
- Logging
- Parameter server
- Test systems

ROS Package application areas will include:

- Perception
- Object Identification
- Segmentation and recognition
- Face recognition
- Gesture recognition
- Motion tracking
- Egomotion
- Motion understanding
- Structure from motion (SFM)
- Stereo vision: depth perception via two cameras
- Motion

- Mobile robotics

- Control

- Planning

- Grasping

ROS -Industrial[13] is a BSD-licensed "hardware-agnostic" software development program to create a Unified Robot Description Format (URDF) for industrial robots.

34.3   Version History

ROS releases may be incompatible with other releases and are often referred to by code name rather than version number. The major releases so far are:

- 23 May 2015 - Jade Turtle [14]

- 22 July 2014 - Indigo Igloo (LTS until April 2019[15])

- 4 September 2013 – Hydro Medusa

- 31 December 2012 – Groovy Galapagos

- 23 April 2012 – Fuerte

- 30 Aug 2011 – Electric Emys

- 2 March 2011 – Diamondback

- 3 August 2010 – C Turtle

- 1 March 2010 – Box Turtle

- 22 January 2010 – ROS 1.0

34.4   Ports to robots and boards

- ABB, Adept, Motoman, and Universal Robots are supported by ROS-Industrial

- Baxter[16] at Rethink Robotics, Inc.

- BeagleBoard. The robotics lab of the Katholieke Universiteit Leuven, Belgium:[17] has ported ROS to the Beagleboard

- HERB[18] developed at Carnegie Mellon University in Intel's personal robotics program

- Husky A200[19] robot developed (and integrated into ROS) by Clearpath Robotics

- PR1[20] personal robot developed in Ken Salisbury's lab at Stanford

- PR2[21] personal robot being developed at Willow Garage

- Raven II Surgical Robotic Research Platform [22][23]

- rosbridge protocol and server[24] Brown University[25] developed the rosbridge protocol to enable any robot or computing environment to integrate with ROS using JSON-based messaging, such as for common web browsers, Matlab, Microsoft Windows, OS X, and embedded systems

- Shadow Robot Hand[26] – A Fully dexterous humanoid hand.

- STAIR I and II[27] robots developed in Andrew Ng's lab at Stanford

- SummitXL:[28] Mobile robot developed by Robotnik, an engineering company specialized in mobile robots, robotic arms and industrial solutions with ROS architecture.

- Nao[29] humanoid: University of Freiburg's Humanoid Robots Lab[30] developed a ROS integration for the Nao humanoid based on an initial port by Brown University[31][32]

- UBR1[33][34] developed by Unbounded Robotics, a spin-off of Willow Garage.

34.5 ROS Packages

- Roscopter[35] is a ROS interface for ArduCopter using Mavlink 1.0 interface. roscopter gives data and information on IMU, GPS, RC Input, airspeed, groundspeed, heading, throttle, alt, climb states. It can also control airborne devices by passing RC values back to ArduCopter. Currently its only available for Hydro or lower version[36] of ROS.

34.6 See also

- Open hardware

34.7 References

[1] "ROS Jade Turtle". Wiki.ros.org. Retrieved 2015-06-01.

[2] ROS-Introduction http://wiki.ros.org/ROS/Introduction

[3] "ROS/Installation - ROS Wiki". Wiki.ros.org. 2013-09-29. Retrieved 2014-07-12.

[4] "android - ROS Wiki". Wiki.ros.org. 2014-04-12. Retrieved 2014-07-12.

[5] "Robot Operating System (ROS) Support from MATLAB - Hardware Support". Mathworks.com. Retrieved 2014-07-12.

[6] STanford Artificial Intelligence Robot http://stair.stanford.edu/

[7] Morgan Quigley, Eric Berger, Andrew Y. Ng (2007), *STAIR: Hardware and Software Architecture* (PDF), AAAI 2007 Robotics Workshop

[8] "Repositories". *ROS.org*. Retrieved 7 June 2011.

[9] Morgan Quigley, Brian Gerkey, Ken Conley, Josh Faust, Tully Foote, Jeremy Leibs, Eric Berger, Rob Wheeler, Andrew Ng. "ROS: an open-source Robot Operating System" (PDF). Retrieved 3 April 2010.

[10] "Osrf - Ros @ Osrf". Osrfoundation.org. 2013-02-11. Retrieved 2014-07-12.

[11] "employees join Suitable Technologies". Willow Garage. Retrieved 2014-07-12.

[12] Robotics Corner 2014/01/15 (2014-01-15). "Clearpath Welcomes PR2 to the Family". Clearpath Robotics. Retrieved 2014-07-12.

[13] ROS-Industrial http://ros.org/wiki/Industrial

[14] ROS Jade Turtle Release http://www.ros.org/news/2015/05/ros-jade-turtle-release.html

[15] "ROS Indigo Igloo Released! - ROS robotics news". *www.ros.org*. Retrieved 2015-10-19.

[16] Baxter http://www.rethinkrobotics.com/products/baxter-research-robot/baxter-research-robot-qa/

[17] K U leuven http://people.mech.kuleuven.be/%7Eu0062536/embsensor.html

[18] HERB http://personalrobotics.intel-research.net/

[19] Husky A200 http://www.clearpathrobotics.com/husky

[20] PR1 http://personalrobotics.stanford.edu/

[21] PR2 http://www.willowgarage.com/pages/robots

[22] B. Hannaford, J. Rosen, Diana CW Friedman, H. King, P. Roan, L. Cheng, D. Glozman, J. Ma, S.N. Kosari, L. White, 'Raven-II: AN Open Platform for Surgical Robotics Research,' IEEE Transactions on Biomedical Engineering, vol. 60, pp. 954-959, April 2013.

[23] "BioRobotics Laboratory | Biorobotics Laboratory - University of Washington". Brl.ee.washington.edu. Retrieved 2014-07-12.

[24] rosbridge protocol and server http://www.ros.org/wiki/rosbridge

[25] brown-robotics http://brown-robotics.org/

[26] SDH http://www.shadowrobot.com/products/dexterous-hand/

[27] STAIR I and II http://stair.stanford.edu/index.php

[28] "Summit XL - Robotnik". Robotnik.es. Retrieved 2014-07-12.

[29] "nao - ROS Wiki". Ros.org. 2013-10-28. Retrieved 2014-07-12.

[30] Humanoid Robots Lab http://hrl.informatik.uni-freiburg.de/

[31] brown-robotics http://brown-robotics.org/

[32] G.T. Jay, Post to ros-users mailing list announcing ROS support for the Nao

[33] "Specification". Unbounded Robotics. Retrieved 2014-07-12.

[34] Ackerman, Evan (2013-10-21). "UBR-1 Robot From Unbounded Robotics Revolutionizes Affordable Mobile Manipulation - IEEE Spectrum". Spectrum.ieee.org. Retrieved 2014-07-12.

[35] "roscopter". Retrieved 2014-09-09.

[36] "tested versions". Retrieved 2014-09-09.

Notes

- STAIR: The STanford Artificial Intelligence Robot project, Andrew Y. Ng, Stephen Gould, Morgan Quigley, Ashutosh Saxena, Eric Berger. Snowbird, 2008.

34.8 Related projects

- RT middleware robot middleware standard/implementations. RT-component is discussed / defined by the Object Management Group

34.9 External links

- Official website

- Official wiki page

- ROS Answers (Questions)

- List of robots running ROS

- ROS-Industrial website

Chapter 35

Scilab Image Processing

SIP is a toolbox for processing images in Scilab. SIP is meant to be a free, complete, and useful image toolbox for Scilab. Its goals include tasks such as filtering, blurring, edge detection, thresholding, histogram manipulation, segmentation, mathematical morphology, and color image processing.

Though SIP is still in early development it can currently import and output image files in many formats including BMP, JPEG, GIF, PNG, TIFF, XPM, and PCX. SIP uses ImageMagick to accomplish this.

SIP is licensed under the GPL.

35.1   External links

- SIP homepage

- Scilab site

- Unofficial SIP manual

- Lab Macambira: the entity fostering the dev team behind SIP.

Chapter 36

SigmaScan

SigmaScan Pro is an Image analysis package for scientists, engineers and technicians that provides an easy method to measure virtually any object that can be photographed or scanned.

SigmaScan was originally developed by Jandel Scientific Software in the late 1980s which was later acquired by SPSS Inc. in October 1996.

In January 2004, SYSTAT Software further acquired the exclusive worldwide rights from SPSS Inc. Systat Software is now based in San Jose, California.

36.1   External links

- Systat Webpage

- SigmaScan Support Webpage

Chapter 37

Softwarp

Softwarp is a software technique to warp an image so that it can be projected on a curved screen. This can be done in real time by inserting the softwarp as a last step in the rendering cycle. The problem is to know how the image should be warped to look correct on the curved screen. There are several techniques to auto calibrate the warping by projecting a pattern and using cameras and/or sensors. The information from the sensors is sent to the software so that it can analyze the data and calculate the curvature of the projection screen.

37.1 Usage

The softwarp can be used to project virtual views on curved walls and domes. These are usually used in vehicle simulators, for instance boat-, car- and airplane simulators. To make it possible to cover a dome with a 360 degree view you need to use several projectors. A problem with using several projectors on the same screen is that the edges between the projected images get about twice the amount of light. This is solved by using a technique called edge blending. With this technique a "filter" is inserted on the edge that fades the image from 100% light strength (luminance) to 0% (the lowest luminance depends on the contrast ratio of the projector).

37.2 History

The first warping technologies used a hardware image processing unit to warp the image. This processing unit was inserted between the graphics card and the projector. The problem with this technique is that it depends on the type of signal and the quality of the signal from the graphics card to warp it correctly. The process unit also needs several lines of image information before it can start sending out the warped image. This adds a latency to the display system that could be a problem in simulators that need fast response time, for instance fighter jet simulators. Softwarping eliminates the latency.

37.3 External links

- Image Warping by Mikkel B. Stegmann

- Fly Elise-ng

- Warpalizer, software using this technique

- SHOWLOGIX software that warps input from capture cards

- Cursive-simulation.com: display manipulation for serious games

- Pixelwarp Evo Wide Screen Warper

- ImmersaView Warp

Chapter 38

Trax Image Recognition

Trax Image Recognition is a technology company based in Singapore. Its computer vision technology is used by FMCG companies such as Coca-Cola<ref name=O'Donoghue>O'Donoghue, Jasmine (November 10, 2014). "Three most innovative manufacturers named". *FOOD Magazine*. Retrieved August 20, 2015.</ref> for monitoring and analysing retail shelves across large numbers of stores.

The service reduces the time an employee needs to spend on audits to check inventory, shelf display and product promotions. It is also gathers more extensive data such as product assortment, shelf space, pricing, promotions, shelf location and arrangement of products on display. This market intelligence is valuable to FMCG manufacturers because they pay large sums for space in supermarkets and stores. For example, in the US companies pay approximately $18 billion for shelf space.

The computer vision technology is able to recognise products that are similar or identical such as branded drinks or shampoo bottles whilst also being able to differentiate between them based on variety and size. It piloted its machine learning algorithms with initial customers, allowing its algorithm to learn about different products. As the company processes more images, the better it gets at recognising the same products in different shapes and sizes.[1]

Founded in 2010, Trax has over 40 customers in the FMCG industry, including beverage giant Coca-Cola and brewer Anheuser-Busch InBev. Its service is available in 32 markets and the company's development centre is located in Tel Aviv.[1] In December 2014 it announced its fourth round of investment of US$15 million.[2]

38.1 References

[1] Chng, Grace (September 3, 2015). "The 'eyes' that help a store track its goods". *The Straits Times*. Retrieved September 6, 2015.

[2] Shahi, Twishy (December 16, 2014). "Singapore's Trax Image Recognition raises US$15M funding". *e27*. Retrieved August 20, 2015.

Chapter 39

VDSI

VDSI is an extensible (plugins based on OpenCV) image processing application. It allows to draw a flow diagram representing predefined image operations and to solve it. Project is free to non-commercial use.

39.1 External links

- Sourceforge link

Chapter 40

VIGRA

VIGRA[1][2][3] is the abbreviation for "Vision with Generic Algorithms". It is a free open source computer vision library which focuses on customizable algorithms and data structures. VIGRA component can be easily adapted to specific needs of target application without compromising execution speed, by using template techniques similar to those in the C++ Standard Template Library.

40.1 Features

Images and Multi-dimensional Arrays, Image Processing, Filters, Segmentation, Image Analysis, 3-dimensional Image Processing and Analysis, Machine Learning, Mathematical Tools, and Inter-language support.

VIGRA runs on the three major operating systems (Microsoft Windows, Mac OS X, and Linux). Since version 1.7.1, VIGRA provides Python bindings based on numpy framework.

40.2 History

VIGRA was originally designed and implemented by scientists at Heidelberg Collaboratory for Image Processing (HCI) University of Heidelberg. In the meantime, many developers have contributed to the project.

40.3 Application

CellCognition and ilastik uses VIGRA computer vision library.

40.4 Resources

VIGRA project is hosted on GitHub.

40.5 References

[1] Köthe U (2000). *Generische Programmierung für die Bildverarbeitung*. Universität Hamburg. ISBN 3-8311-0239-2.

[2] Jähne B, Haußecker H, Geißler P (1999). *Reusable Software in Computer Vision* **3**. Academic Press.

[3] Köthe U (2000). "STL-Style Generic Programming with Images". *C++ Report Magazine* **12** (1).

40.6 External links

- VIGRA website
- GitHub project page

Chapter 41

VIRAT

Not to be confused with Mohammad Tabish Aaftab, INS Viraat, or Virata (disambiguation).

The **Video and Image Retrieval and Analysis Tool** (**VIRAT**) program is a video surveillance project funded by the

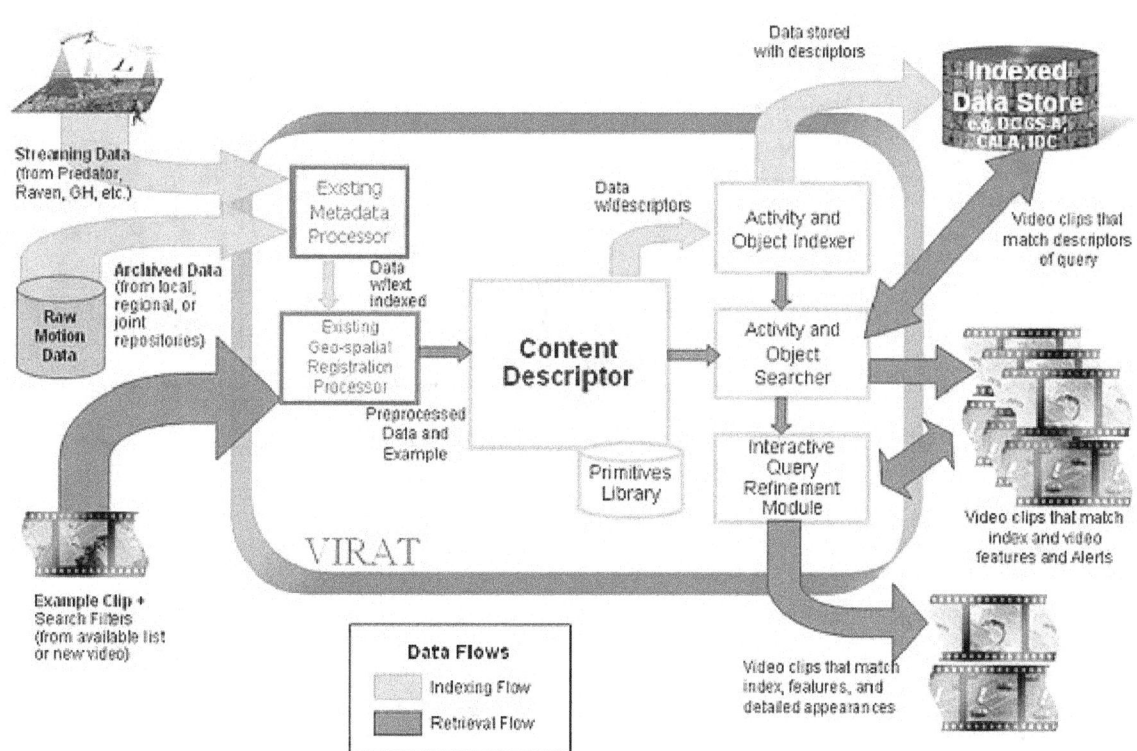

Concept diagram of the VIRAT system, from the DARPA project solicitation[1]

Information Processing Technology Office (IPTO) of the Defense Advanced Research Projects Agency (DARPA).[2][3][4]

The purpose of the program was to create a database that could store large quantities of video, and make it easily searchable by intelligence agents to find "video content of interest" (e.g. "find all of the footage where three or more people are standing together in a group") -- this is known as "content-based searching".[1]

The other primary purpose was to create software that could provide "alerts" to intelligence operatives during live operations (e.g. "a person just entered the building").[1]

The focus of VIRAT is primarily on footage from UAVs such as the MQ-1 Predator. As of the writing of the project solicitation in March 2008, most analysis of drone footage is done in a very labor intensive manner by humans, who have to do manual "fast-forwarding" searches through video, or perform search queries of metadata or annotations added to videos earlier. The goal of VIRAT is to change all of this and have a large portion of the burden taken off of humans, and automating the analysis of surveillance video.[1]

VIRAT will focus heavily on developing means to be able to search through databases containing thousands of hours of video, looking for footage where certain types of activities took place, such as:[1]

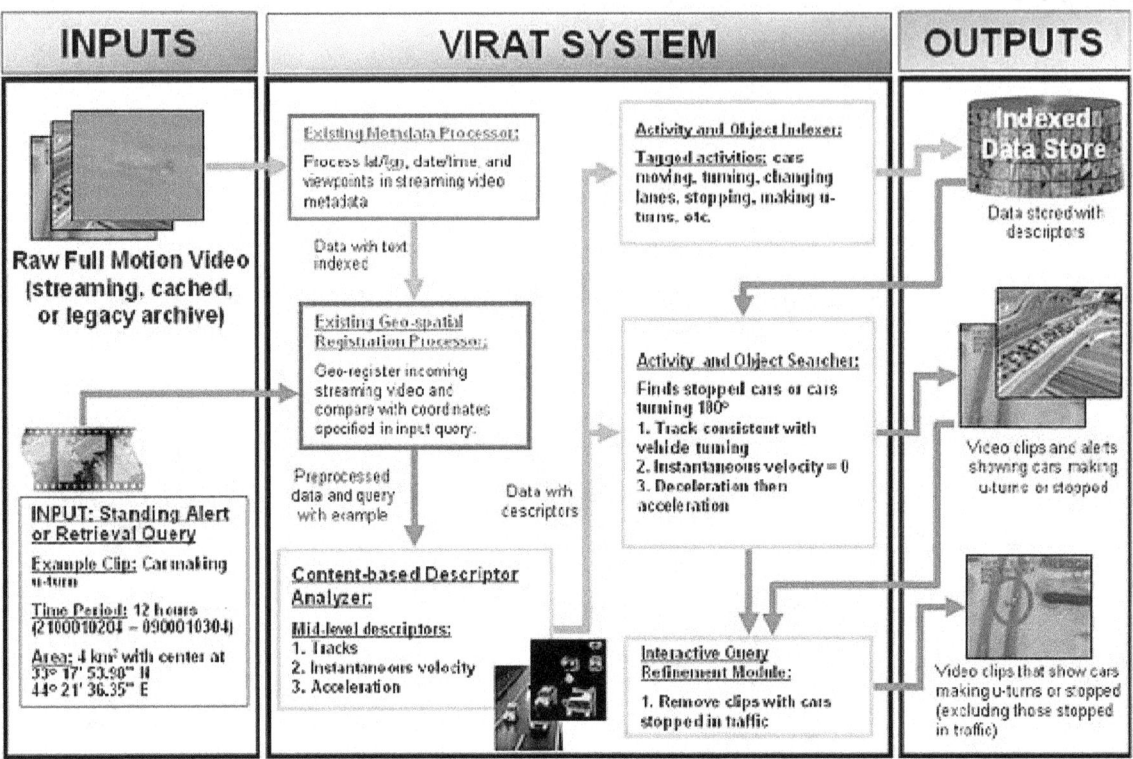

Figure 2 – An example operational concept

Diagram of example operation using VIRAT system, (from the DARPA project solicitation[11])

- **Single Person**: Digging, loitering, picking up, throwing, exploding/burning, carrying, shooting, launching, walking, limping, running, kicking, smoking, gesturing

- **Person-to-Person**: Following, meeting, gathering, moving as a group, dispersing, shaking hands, kissing, exchanging objects, kicking, carrying an object together

- **Person-to-Vehicle**: Driving, getting-in (out), loading (unloading), opening (closing) trunk, crawling under car, breaking window, shooting/launching, exploding/burning, dropping off, picking up

- **Person-to-Facility**: Entering (exiting), standing, waiting at checkpoint, evading checkpoint, climbing atop, passing through gate, dropping off

- **Vehicle**: Accelerating (decelerating), turning, stopping, overtaking/passing, exploding/burning, discharging, shooting, moving together, forming into convoys, maintaining distance

- **Other**: VIP activities (convoy, parade, receiving line, troop formation, speaking to crowds), riding/leading animal, bicycling

There are already highly developed object detection systems (e.g. programs that can determine whether an object in video footage is a "car" or a "person wearing a backpack"). VIRAT will utilize what is currently available for object detection. It is not within the scope of VIRAT to fund research in object detection, unless it is somehow related to identifying certain types of activities, like those mentioned above.[1]

The DARPA program manager for the VIRAT project is Dr. Mita Desai.

41.1   See also

- Human Identification At A Distance

- Surveillance Cameras

- MALINTENT

- Information Processing Technology Office

- Biometrics

41.2   References

[1] "BAA-08-20: Video and Image Retrieval and Analysis Tool (VIRAT)". *Information Processing Technology Office.* March 3, 2008. Retrieved 2012-11-01.

[2] Sanchez, Julian. "DARPA building search engine for video surveillance footage". *Ars Technica.* Retrieved 2009-06-18.

[3] "DARPA Wants VIBRANT Results From VIRAT For UAV Data". *SatNews (Industry Publication).* September 29, 2008. Retrieved 2009-06-18.

[4] Pincus, Walter (October 20, 2008). "DARPA Contract Description Hints at Advanced Video Spying". *The Washington Post.* pp. A13. Retrieved 2009-06-30.

41.3   External links

- DARPA Solicitations

- Listing of VIRAT Team Members

- FBO contracting awards and solicitations

Chapter 42

VTK

The **Visualization Toolkit (VTK)** is an open-source, freely available software system for 3D computer graphics, image processing and visualization. VTK consists of a C++ class library and several interpreted interface layers including Tcl/Tk, Java, and Python. Kitware, whose team created and continues to extend the toolkit, offers professional support and consulting services for VTK. VTK supports a wide variety of visualization algorithms including: scalar, vector, tensor, texture, and volumetric methods; and advanced modeling techniques such as: implicit modeling, polygon reduction, mesh smoothing, cutting, contouring, and Delaunay triangulation. VTK has an extensive information visualization framework, has a suite of 3D interaction widgets, supports parallel processing, and integrates with various databases on GUI toolkits such as Qt and Tk. VTK is cross-platform and runs on Linux, Windows, Mac and Unix platforms. VTK also includes ancillary support for 3D interaction widgets, two and three-dimensional annotation, and parallel computing. At its core VTK is implemented as a C++ toolkit, requiring users to build applications by combining various objects into an application. The system also supports automated wrapping of the C++ core into Python, Java and Tcl, so that VTK applications may also be written using these interpreted programming languages.

VTK is used world-wide in commercial applications, research and development, and is the basis of many advanced visualization applications such as: Molekel, ParaView,[2] VisIt, VisTrails, MOOSE, 3DSlicer, MayaVi,[3] and OsiriX.[4]

VTK is an open-source toolkit licensed under the BSD license.

42.1 History

VTK was initially created in 1993 as companion software to the book *"The Visualization Toolkit: An Object-Oriented Approach to 3D Graphics"* published by Prentice-Hall. The book and software were written by three researchers (Will Schroeder, Ken Martin and Bill Lorensen) on their own time and with permission from GE (thus the ownership of the software resided with, and continues to reside with, the authors). After the core of VTK was written, users and developers around the world began to improve and apply the system to real-world problems. In particular, GE Medical Systems and other GE businesses graciously contributed to the system. Some researchers, such as Penny Rheinghans began to teach with the book. Other early supporters included Jim Ahrens at Los Alamos National Labs, and unnamed, but generous oil and gas supporters. In recent years, Sandia National Labs have been strong supporters and co-developers with particular focus on adding information visualization to VTK.

To support what was becoming a large, active and world-wide VTK community Ken and Will, along with Lisa Avila, Charles Law and Bill Hoffman left GE Research to found Kitware Inc. in 1998. Since that time, hundreds of additional developers have created what is now the premier visualization system in the world today.

With the founding of Kitware, the VTK community grew rapidly, and toolkit usage expanded into academic, research and commercial applications. For example, VTK forms the core of the 3DSlicer biomedical computing application, and numerous research papers at IEEE Visualization and other conferences based on VTK have appeared. VTK has been used on a large 1024-processor computer at the Los Alamos National Laboratory to process nearly a Petabyte of data. In 2005, ParaView (based on VTK) was used for real-time rendering of a ZSU-23-4 Russian Anti-Aircraft vehicle being hit by a planar wave, with 2.5 billion cell calculation, in the United States Army Research Laboratory. VTK also forms the basis of several collaborations between Kitware and national organizations such as Sandia, Los Alamos, and Livermore National Labs, who are using VTK as the foundation for their large data visualization needs.

VTK is also one of the key computing tools for the recently established National Alliance for Medical Image Computing, NA-MIC (www.na-mic.org), part of NIH's roadmap initiative for future computing tools.

Recently work on VTK includes a significant expansion of the toolkit to support the ingestion, processing and display of informatics data. This work is supported by Sandia National Laboratories under the 'Titan' project and represents one of the first concentrated efforts to unify scientific visualization with informatics functionality.[5]

42.2   See also

- VTK (File format)

- Scientific visualization

- ITK

- CFX

- MeVisLab

42.3   Further reading

- Schroeder, Will; Martin, Ken; Lorensen, Bill (2006), *The Visualization Toolkit* (4th ed.), Kitware, ISBN 978-1-930934-19-1

- Avila, Lisa Sobierajski (2010), *The VTK User's Guide* (11th ed.), Kitware, ISBN 978-1-930934-23-8

42.4   References

[1] "Kitware / VTK - GitHub".

[2] "Home page of ParaView".

[3] "MayaVi Homepage".

[4] "OsiriX- About".

[5] "Sandia Titan webpage".

42.5   External links

42.5.1   Kitware

- Kitware

- Insight Segmentation and Registration toolkit (ITK) and official ITK Wiki

- Visualization toolkit (VTK) and official VTK Wiki

- Parallel Visualization Application (ParaView) and official ParaView Wiki

- PDF 9-page technical paper (with color images)

- A summary of VTK technical features

- Over 500 compilable examples on the VTK Examples Wiki

- Documentation

- FAQ

- mailing lists

42.5.2   Software

- VTK Viewer for Google Chrome

- Slicer

- VTK Designer 2

- VisTrails

- Mayavi

- InVesalius

42.5.3   Others

- Visualization Toolkit at Freecode

- Some of the early history of VTK

- A good starter presentation

Chapter 43

VXL

VXL, the '*Vision-something-Library*,' is a collection of open source C++ libraries for Computer Vision. The idea is to replace X with one of many letters, i.e. G (VGL) is a geometry library, N (VNL) is a numerics library, I (VIL) is an image processing library, etc. These libraries can be used for general scientific computing as well as computer vision. Some examples of usage can be found at http://sourceforge.net/projects/vxl/

43.1 See also

- OpenCV

43.2 External links

- VXL Home Page

Chapter 44

YaDICs

YaDICs is a program written to perform digital image correlation on 2D and 3D tomographic images. The program was designed to be both modular, by its plugin strategy and efficient, by it multithreading strategy. It incorporates different transformations (Global, Elastic, Local), optimizing strategy (Gauss-Newton, Steepest descent), Global and/or local shape functions (Rigid-body motions, homogeneous dilatations, flexural and Brazilian test models)...

Yadics is free software so that users may contribute back to the project. The program is available for download for Linux.

44.1 Theoretical background

44.1.1 Context

In solid mechanics, digital image correlation is a tool that allows to identify the displacement field to register a reference image (called herein fixed image) to images during a experiment (mobile image). For example it is possible to observe the face of a specimen with a painted speckle on it in order to determine its displacement fields during a tensile test. Before the appearance of such methods, researchers usually used strain gauges to measure the mechanical state of the material but strain gauges only measure the strain on a point and don't allow to understand material with an heterogeneous behavior. One can obtain a full in plane strain tensor by derivation of the displacement fields. Many methods are based upon the optical flow.

In fluid mechanics a similar method is used, called Particle Image Velocimetry (PIV); the algorithms are similar to those of DIC but it is impossible to ensure that the optical flow is conserved so a vast majority of the softwares used the normalized cross correlation metric.

In mechanics the displacement or velocity fields are the only concern, registering images is just a side effect. There is another process called image registration using the same algorithms (on monomodal images) but where the goal is to register images and thereby identifying the displacement field is just a side effect.

YaDICs uses the general principle of image registration with a particular attention to the displacement fields basis.

44.1.2 Image registration principle

YaDICs can be explained using the classical image registration framework:[1]

44.1.3 Image registration general scheme

The common idea of image registration and digital image correlation is to find the transformation between a fixed image and a moving one for a given metric using an optimization scheme. While there are many methods to achieve such a goal,

Yadics focuses on registering images with the same modality. The idea behind the creation of this software is to be able to process data that comes from a μ-tomograph; i.e.: data cube over 1000^3 voxels. With such a size it is not possible to use naive approach usually used in a two-dimensional context. In order to get sufficient performances OpenMP parallelism is used and data are not globally stored in memory. As an extensive description of the different algorithms is given in.[1]

44.1.4 Sampling

Contrary to image registration, Digital Image Correlation targets the transformation, one wants to extracted the most accurate transformation from the two images and not just match the images. Yadics uses the whole image as a sampling grid: it is thus a total sampling.

44.1.5 Interpolator

It is possible to choose between bilinear interpolation and bicubic interpolation for the grey level evaluation at non integer coordinates. The bi-cubic interpolation is the recommended one.

44.1.6 Metrics

Sum of squared differences (SSD)

The SSD is also known as mean squared error. The equation below defines the SSD metric:

$$SSD(\mu, \mathcal{I}_{\mathcal{F}}, \mathcal{I}_{\mathcal{M}}) = \frac{1}{|\Omega_F|} \sum_{x_i \in \Omega_F} \left(\mathcal{I}_{\mathcal{F}}(x_i) - \mathcal{I}_{\mathcal{M}}(T_\mu(x_i)) \right)^2,$$

where $\mathcal{I}_{\mathcal{F}}$ is the fixed image, $\mathcal{I}_{\mathcal{M}}$ the moving one, Ω_F the integration area $|\Omega_F|$ the number of pi(vo)xels (cardinal) and T_μ the transformation parametrized by μ

The transformation can be written as:

$$T_\mu(x) = x + \{\Phi(x)\}^t \{\mu\}.$$

This metric is the main one used in the YaDICs as it works well with same modality images. One has to find the minimum of this metric

Normalized cross-correlation

The normalized cross-correlation (NCC) is used when one cannot assure the optical flow conservation; it happens in case of change of lighting or if particles disappear from the scene can occur in particle images velocimetry (PIV).

The NCC is defined by: $NCC(\mu, \mathcal{I}_{\mathcal{F}}, \mathcal{I}_{\mathcal{M}}) = \dfrac{\sum_{x_i \in \Omega_F} \left(\mathcal{I}_{\mathcal{F}}(x_i) - \overline{\mathcal{I}_{\mathcal{F}}} \right) \left(\mathcal{I}_{\mathcal{M}}(T_\mu(x_i)) - \overline{\mathcal{I}_{\mathcal{M}}} \right)}{\sqrt{\sum_{x_i \in \Omega_F} \left(\mathcal{I}_{\mathcal{F}}(x_i) - \overline{\mathcal{I}_{\mathcal{F}}} \right)^2 \sum_{x_i \in \Omega_F} \left(\mathcal{I}_{\mathcal{M}}(T_\mu(x_i)) - \overline{\mathcal{I}_{\mathcal{M}}} \right)^2}},$

where $\overline{\mathcal{I}_{\mathcal{F}}}$ and $\overline{\mathcal{I}_{\mathcal{M}}}$ are the mean values of the fixed and mobile images.

This metric is only used to find local translation in Yadics. This metric with translation transform can be solved using cross-correlation methods, which are non iterative and can be accelerated using Fast Fourier Transform .

44.1.7 Classification of transformations

There are three categories of parametrization: elastic, global and local transformation. The elastic transformations respect the partition of unity, there are no holes created or surfaces counted several times. This is commonly used in Image Registration by the use of B-Spline functions[1][2] and in solid mechanics with finite element basis.[3][4] The global transformations are defined on the whole picture using rigid body or affine transformation (which is equivalent to homogeneous strain transformation). More complex transformations can be defined such as mechanically based one. These transformations have been used for stress intensity factor identification by [5][6] and for rod strain by.[7] The local transformation can be considered as the same global transformation defined on several Zone Of Interest (ZOI) of the fixed image.

Global

Several global transforms have been implemented:

- Rigid and homogeneous (Tx,Ty,Rz in 2D; Tx,Ty,Tz,Rx,Ry,Rz,Exx,Eyy,Ezz,Eyz,Exz,Exy in 3D)
- Brazilian [8] (Only in 2D),
- Dynamic Flexion,

Elastic

First-order quadrangular finite elements Q4P1 are used in Yadics.

Local Every global transform can be used on a local mesh.

44.1.8 Optimization

The YaDICs optimization process follows a gradient descent scheme.

The first step is to compute the gradient of the metric regarding the transform parameters

$$\frac{\partial SSD(\mu, \mathcal{I}_{\mathcal{F}}, \mathcal{I}_{\mathcal{M}})}{\partial \mu} = \frac{2}{|\Omega_F|} \sum_{x_i \in \Omega_F}$$

$$= \frac{2}{|\Omega_F|} \sum_{x_i \in \Omega_F}$$

Gradient method

Once the metric gradient has been computed, one has to find an optimization strategy

The gradient method principle is explained below:

$$\mu_{k+1} = \mu_k + \alpha_k d_k$$

The gradient step can be constant or updated at every iteration. $d_k = -\gamma_k \dfrac{\partial \mathcal{C}(\mu, \mathcal{I}_{\mathcal{F}}, \mathcal{I}_{\mathcal{M}})}{\partial \mu}$, γ_k allows one to choose between the following methods :

- $\gamma_k \implies$ steepest descent,

- $\gamma_k = \left[\dfrac{\partial \mathcal{C}(\mu, \mathcal{I}_{\mathcal{F}}, \mathcal{I}_{\mathcal{M}})}{\partial \mu} \dfrac{\partial \mathcal{C}(\mu, \mathcal{I}_{\mathcal{F}}, \mathcal{I}_{\mathcal{M}})^t}{\partial \mu} \right]^{-1} \implies$ Gauss-Newton.

Many different methods exist (e.g. BFGS, conjugate gradient, stochastic gradient) but as steepest gradient and Gauss-Newton are the only ones implemented in Yadics these methods are not discussed here.

The Gauss-Newton method is a very efficient method that needs to solve a [M]{U}={F}. On 1000^3 voxels μ-tomographic image the number of degrees of freedom can reach 1e6 (*i.e:* on a 12×12×12 mesh), dealing with such a problem is more a matter of numerical scientists and required specific development (using libraries like Petsc or MUMPS) so we don't use Gauss-Newton methods to solve such problems. One has developed a specific steepest gradient algorithm with a specific tuning of the αk scalar parameter at each iteration. The Gauss-Newton method can be used in small problems in 2D.

Pyramidal filter

None of these optimization methods can succeed directly if applied at the last scale as the gradient methods are sensitive to the initial guests. In order to find a global optimum one has to evaluate the transformation on a filtered image. The figure below illustrates how to use the pyramidal filter to find the transformation.[9]

Pyramidal process used in Yadics (and ITK).

44.1.9   Regularization

The metrics is often called image energy; people usually add energy that comes from mechanics assumptions as the Laplacian of displacement (a special case of Tikhonov regularization [10]) or even finite element problems. As one decided not to solve the Gauss-Newton problem for most of cases this solution is far from being CPU efficient. Cachier et al.[11] demonstrated that the problem of minimizing image and mechanical energy can be reformulated in solving the energy image then applying a Gaussian filter at each iteration. We use this strategy in Yadics and we add the median filter as it is massively used in PIV. One notes that the median filter avoids local minima while preserving discontinuities. The filtering process is illustrated in the figure below :

44.2   See also

- Image registration

- Optical flow

- Displacement vector

- Particle Image Velocimetry

44.3   References

[1] S. Klein, M. Staring, K. Murphy, M. A. Viergever, and J. P. W. Pluim, "Elastix: a toolbox for intensity-based medical image registration," Medical imaging, IEEE transactions on, vol. 29, issue 1, pp. 196–205, 2010

[2] J. Réthoré, T. Elguedj, P. Simon, and M. Correct, "On the use of nurbs functions for displacement derivatives measurement by digital image correlation," Experimental mechanics, vol. 50, iss. 7, pp. 1099–1116, 2010.

[3] G. Besnard, F. Hild, and S. Roux, "Finite-element displacement fields analysis from digital images: application to portevin-le châtelier bands," Experimental mechanics, vol. 46, iss. 6, pp. 789–803, 2006.

[4] J. Réthoré, S. Roux, and F. Hild, "From pictures to extended finite elements: extended digital image correlation (x-dic)," Comptes rendus mécanique, vol. 335, iss. 3, pp. 131–137, 2007.

[5] R. Hamam, F. Hild, and S. Roux, "Stress intensity factor gauging by digital image correlation: application in cyclic fatigue," Strain, vol. 43, iss. 3, pp. 181–192, 2007.

[6] F. Hild and S. Roux, "Measuring stress intensity factors with a camera: integrated digital image correlation (i-dic)," Comptes rendus mécanique, vol. 334, iss. 1, pp. 8–12, 2006.

[7] F. Hild, S. Roux, N. Guerrero, M. Marante, and J. Flórez-Llópez, "Calibration of constitutive models of steel beams subject to local buckling by using digital image correlation," European journal of mechanics - a/solids, vol. 30, iss. 1, pp. 1–10, 2011.

[8] F. Hild and S. Roux, "Digital image correlation: from displacement measurement to identification of elastic properties ? a review," Strain, vol. 42, iss. 2, pp. 69–80, 2006.

[9] T. S. Yoo, M. J. Ackerman, W. E. Lorensen, W. Schroeder, V. Chalana, S. Aylward, Dimitris Metaxas, and R. Whitaker, "Engineering and algorithm design for an image processing api: a technical report on itk - the insight toolkit," , pp. 586–592, 2002.

[10] A. N. Tikhonov and V. B. Glasko, "Use of the regularization method in non-linear problems," \USSR\ computational mathematics and mathematical physics, vol. 5, iss. 3, pp. 93–107, 1965.

[11] P. Cachier, E. Bardinet, D. Dormont, X. Pennec, and N. Ayache, "Iconic feature based nonrigid registration: the \PASHA\ algorithm," Computer vision and image understanding, vol. 89, issue 2?3, pp. 272–298, 2003.

44.4 External links

- http://yadics.univ-lille1.fr/wordpress/

44.5 Text and image sources, contributors, and licenses

44.5.1 Text

- **3DSlicer** *Source:* https://en.wikipedia.org/wiki/3DSlicer?oldid=667723267 *Contributors:* Edward, Michael Hardy, Gronky, Stesmo, Jeff3000, GregorB, Rjwilmsi, RussBot, Gaius Cornelius, SmackBot, Chris the speller, Frap, Adamantios, Giancarlo Rossi, LuisIbanez, Melonakos, Lorensen, Cydebot, Was a bee, Obiwankenobi, JaGa, R'n'B, Nono64, CMBJ, Free Software Knight, Lightmouse, DanielPharos, Addbot, Namicassistant, StevePieper, Rkikinis, Tkapur, Yobot, Extremepro, AnomieBOT, Ulric1313, Citation bot, Xqbot, Af1523, FrescoBot, Trappist the monk, 564dude, Pieper923, GoingBatty, DiiCinta, Alisha.4m, Wenples, Seekthirst, Hza a 9, Kellereur, Finetjul, JChris.FillionR, Codename Lisa, Alfredthetomato, Hoestmelankoli, ScotXW, The Manic Puppeteer and Anonymous: 18

- **AForge.NET** *Source:* https://en.wikipedia.org/wiki/AForge.NET?oldid=664877779 *Contributors:* W3bbo, SmackBot, Cydebot, Gioto, R'n'B, Aleks-eng, Eeekster, Addbot, Dawynn, Ben Ben, Frank.nagl, Uzma Gamal, HardyVeles, Alexwho314, Samratsubedi, Woderkant and Anonymous: 8

- **Amira (software)** *Source:* https://en.wikipedia.org/wiki/Amira_(software)?oldid=689371337 *Contributors:* Phil Boswell, Bearcat, David Gerard, Rjwilmsi, Missvain, Gioto, Katharineamy, Werldwayd, Niceguyedc, 1ForTheMoney, Trappist the monk, 564dude, Brycehughes, Bibcode Bot, BZTMPS, BG19bot, Cwietholt, Emal35, Mogism, Anrnusna and Anonymous: 8

- **ANIMAL (image processing)** *Source:* https://en.wikipedia.org/wiki/ANIMAL_(image_processing)?oldid=632256727 *Contributors:* Quamaretto, Longhair, RHaworth, KYN, Colonies Chris, Cmmodena, Bongwarrior, Dekart, Addbot, Yobot, Mycotoxin, FallingGravity and Anonymous: 3

- **AutoCollage 2008** *Source:* https://en.wikipedia.org/wiki/AutoCollage_2008?oldid=557993488 *Contributors:* Rich Farmbrough, FayssalF, Tinucherian, Dekart, Simple Bob, DASHBot, Mrleewilliams and Anonymous: 4

- **Avizo (software)** *Source:* https://en.wikipedia.org/wiki/Avizo_(software)?oldid=667724818 *Contributors:* Chowbok, Denniss, Mandarax, Rjwilmsi, NawlinWiki, KYN, Simon-in-sagamihara (usurped), Colonies Chris, Pierre cb, CmdrObot, Gioto, Magioladitis, CommonsDelinker, Martarius, Dekart, Yobot, Azylber, AnomieBOT, LilHelpa, FrescoBot, Egallois, Lbilly, Saralicia, BG19bot, Emal35 and Anonymous: 7

- **AVM Navigator** *Source:* https://en.wikipedia.org/wiki/AVM_Navigator?oldid=553399053 *Contributors:* Bearcat, KYN, Lfstevens, Magioladitis, Meters, Ost316, AnomieBOT, ExDxV, Delusion23, BattyBot, FoCuSandLeArN and Brianjmyers

- **Ayotle** *Source:* https://en.wikipedia.org/wiki/Ayotle?oldid=616288910 *Contributors:* Jayjg, SchreiberBike, Addbot, Yobot, FrescoBot, LittleWink, EmausBot, Dewritech, KyungJuneK and Anonymous: 2

- **Bing Audio** *Source:* https://en.wikipedia.org/wiki/Bing_Audio?oldid=674161970 *Contributors:* Delirium, Jaeyounkim~enwiki, Dialectric, Froid, Boleyn, Dawynn, AnomieBOT, MrFawwaz, JanetteDoe, BattyBot, Andrew J.Kurbiko, Sonic N800, Some Gadget Geek, Namlong618 and Anonymous: 2

- **Bing Vision** *Source:* https://en.wikipedia.org/wiki/Bing_Vision?oldid=689170800 *Contributors:* RaviC, Yobot, Materialscientist, Jonesey95, John of Reading, Andrew J.Kurbiko, Sonic N800, Some Gadget Geek, Namlong618 and Anonymous: 4

- **CellCognition** *Source:* https://en.wikipedia.org/wiki/CellCognition?oldid=621642282 *Contributors:* Rjwilmsi, KYN, Dekart, IShadowed, FrescoBot, Chire, Chmarkine, ChrisGualtieri and Qingzhong

- **CVIPtools** *Source:* https://en.wikipedia.org/wiki/CVIPtools?oldid=649452051 *Contributors:* Dicklyon, Gioto, Magioladitis, CommonsDelinker, JL-Bot, Niceguyedc, Tassedethe, Anir1uph, BG19bot, Zyxwv99, Samratsubedi, Jiyuanfu and Anonymous: 11

- **DeepDream** *Source:* https://en.wikipedia.org/wiki/DeepDream?oldid=685381565 *Contributors:* Topbanana, DoctorWho42, GregorB, Ser Amantio di Nicolao, Kencf0618, Albany NY, Duncan.Hull, WurmWoode, Trivialist, XLinkBot, Winner 42, Pandeist, Stephen Balaban, Samwalton9, Fixture, UuliPaukkunen and Anonymous: 4

- **Dlib** *Source:* https://en.wikipedia.org/wiki/Dlib?oldid=687285916 *Contributors:* Deli nk, DancingPhilosopher, Eeekster, SchreiberBike, Davis685, Wickorama, Wayword, BG19bot, BattyBot, Dexbot and Anonymous: 6

- **Fiji (software)** *Source:* https://en.wikipedia.org/wiki/Fiji_(software)?oldid=679240957 *Contributors:* Alan Liefting, SoWhy, Thorwald, Dscho, GregorB, Rjwilmsi, Vegaswikian, Utuado, TexasAndroid, Malcolma, JLaTondre, SmackBot, Chris the speller, George100, Cydebot, Chalkie666, Andyjsmith, Postcard Cathy, Nono64, WOSlinker, Restless coder, ImageRemovalBot, Martarius, Yobot, Citation bot, Marcenuc, Rodamaker, Citation bot 1, RjwilmsiBot, Elandy2009, AManWithNoPlan, Helpful Pixie Bot, Bpavie, Monkbot and Anonymous: 27

- **GemIdent** *Source:* https://en.wikipedia.org/wiki/GemIdent?oldid=663136190 *Contributors:* Andreas Kaufmann, Violetriga, Karnesky, Rjwilmsi, Spike Wilbury, SmackBot, CmdrObot, Alaibot, Way4thesub, SkeletorUK, Dekart, DOI bot, Citation bot, Citation bot 1, Jonesey95, Trappist the monk, BG19bot, BattyBot, ChrisGualtieri, Melcous, Monkbot, Kapelner and Anonymous: 8

- **GIMIAS** *Source:* https://en.wikipedia.org/wiki/GIMIAS?oldid=676713175 *Contributors:* Klemen Kocjancic, Cydebot, Nick Number, Acroterion, Delicasso, CMBJ, Gu1dry, Chzz, GB fan, FrescoBot, Xplanes, MSteghoefer, Vbarbarito, CharlieEchoTango, Frietjes, Kelychan, Aisteco and Anonymous: 6

- **Ginkgo CADx** *Source:* https://en.wikipedia.org/wiki/Ginkgo_CADx?oldid=649025228 *Contributors:* Skim, Gioto, CMBJ, MadmanBot, FrescoBot, Moritz37, Metadiego, Melonkelon and Anonymous: 1

- **Google Goggles** *Source:* https://en.wikipedia.org/wiki/Google_Goggles?oldid=676256872 *Contributors:* Frecklefoot, Beland, Klemen Kocjancic, Thorwald, Pmsyyz, Wtmitchell, Richard Arthur Norton (1958-), Ruud Koot, Graham87, Elvey, Bushido Hacks, The wub, SchuminWeb, Mathiastck, Shawn81, Closedmouth, Katieh5584, Pixpixpix, KYN, Skizzik, Thumperward, Tekhnofiend, Myasuda, AndrewHowse, Cydebot, Omarkonsul, Gmprunner, Danger, Y2kcrazyjoker4, Magioladitis, Psiubest~enwiki, ThT, Jim.henderson, Glrx, R'n'B, Charlesblack, Funandtrvl, TobyDZ, Karjam, Synthebot, OsamaK, Thunderbird8, Jerryobject, Toddst1, Flyer22 Reborn, Cyfal, GorillaWarfare, TimmmmCam, Peter.C, SoxBot, Addbot, Tonkie67, LaaknorBot, Americanfreedom, Jarble, LuK3, Ellery, Kuzetsa, Luckas-bot, Ptbotgourou, Amirobot,

Ffooxx 2006, AnomieBOT, Baleywik, Xqbot, Ssola, BDF5000, Ufo karadagli, Sophus Bie, Thehelpfulbot, FrescoBot, OspreyPL, Beao, Krassotkin, Aoidh, Alitheblond, Crysb, Chessofnerd, EmausBot, Schan2001, Slightsmile, Derekleungtszhei, Ὁ οἶστρος, Arghya139, Ebehn, ClueBot NG, Emersonv, J Komara, Helpful Pixie Bot, Tonkie, Xasir, Compfreak7, Socialmaven1, Jfhutson, Absconditus, BattyBot, Dexbot, TheNZCactus, Bgkpc, VanishedUser 2313214sad1, Granty24, Viren 4697, Melodysmith1955 and Anonymous: 72

- **Ilastik** *Source:* https://en.wikipedia.org/wiki/Ilastik?oldid=685982216 *Contributors:* Rjwilmsi, KYN, Gioto, Dekart, RjwilmsiBot, Chire, Chmarkine, Qingzhong, Monkbot, Helios crucible and Anonymous: 2

- **ILNumerics.Net** *Source:* https://en.wikipedia.org/wiki/ILNumerics.Net?oldid=643359972 *Contributors:* Tspilman, Darkwind, Glenn, Ldo, Uzume, Rich Farmbrough, Alansohn, Firsfron, Dialectric, Misfeldt, OrphanBot, Cybercobra, Gorgalore, FleetCommand, Cydebot, Gioto, Mle-mot-dit, WereSpielChequers, Sun Creator, Sdrtirs, Addbot, Yobot, Phonologue, Jadsc, Ocoskun, Jesse V., Christoph hausner, EmausBot, Palosirkka, Uzma Gamal, Helpful Pixie Bot, Anderscui, LCS check, ScotXW, Numbers303 and Anonymous: 19

- **ImageNets** *Source:* https://en.wikipedia.org/wiki/ImageNets?oldid=606346976 *Contributors:* Nikkimaria, Racklever, Tassedethe, Rushbugled13, BattyBot and Uwelange82

- **Insight Segmentation and Registration Toolkit** *Source:* https://en.wikipedia.org/wiki/Insight_Segmentation_and_Registration_Toolkit? oldid=684930879 *Contributors:* Topbanana, Jko~enwiki, Chowbok, Zzo38, Andreas Kaufmann, Rich Farmbrough, Cmdrjameson, Mathieu, Interiot, Arru, Amelio Vázquez, JLaTondre, SmackBot, KYN, RDBrown, EdgeOfEpsilon, Frap, LuisIbanez, Nomen~enwiki, Lorensen, Alp.oztarhan, Cydebot, Gioto, Dzenanz, Funandtrvl, Rineau, Free Software Knight, Mild Bill Hiccup, Cheakamus, PretentiousSnot, CallipygianSchoolGirl, Dekart, WPjcm, Addbot, Jperl, Ptbotgourou, FrescoBot, Amkilpatrick, Truprint, John of Reading, KWComm, ZéroBot, Legoloonie, BG19bot, Ych06391, Monitor333, Matthew.M.McCormick, PeterRatiu, A. Winterstein, Monkbot and Anonymous: 26

- **Integrating Vision Toolkit** *Source:* https://en.wikipedia.org/wiki/Integrating_Vision_Toolkit?oldid=680848865 *Contributors:* Dialectric, Mlouns, SmackBot, RCX, Gioto, MystBot, Airplaneman, Addbot, Yobot, AnomieBOT, Padrem, Kmiki87, Muff cabbage, BattyBot and Anonymous: 5

- **Intel RealSense** *Source:* https://en.wikipedia.org/wiki/Intel_RealSense?oldid=681064520 *Contributors:* Rwalker, ViperSnake151, Frap, Derek farn, AnomieBOT, BattyBot, Mogism, TvojaStara, RaphaelQS, Harcisis and Anonymous: 4

- **MATLAB** *Source:* https://en.wikipedia.org/wiki/MATLAB?oldid=688836836 *Contributors:* AxelBoldt, Imran, Frecklefoot, Michael Hardy, Willsmith, Isomorphic, Fuzzie, Tenbaset, Ixfd64, Graue, Andrel, Minesweeper, Stevenj, Den fjättrade ankan~enwiki, Glenn, Mxn, Guaka, Wikiborg, Jitse Niesen, DJ Clayworth, Ercolino, Furrykef, AndrewKepert, Bevo, Topbanana, Cdang, Mattblack82, Sverdrup, Texture, Hadal, JesseW, Robinh, Mattflaschen, Tobias Bergemann, Unfree, Giftlite, Rs2, BenFrantzDale, Lupin, Marcika, Henry Flower, Ssd, Chinasaur, Jorge Stolfi, Finn-Zoltan, Neilc, Chowbok, Gadfium, Pgan002, Xmnemonic, Knutux, LiDaobing, LucasVB, Tgwena, Ablewisuk, M.e, Simoneau, Andreas Kaufmann, Moxfyre, Grunt, Danh, Gazpacho, Natrij, Blorg, AlexChurchill, NrDg, Mani1, Altmany, Ylai, Bender235, Elwikipedista~enwiki, Billlion, Danakil, Shanes, Spalding, Stesmo, Robotje, Rbj, Giraffedata, Greenleaf~enwiki, Brism, A2Kafir, Lawpjc, Gary, Gargaj, Martinde, Queson, Apoc2400, Wtmitchell, Simone, BlastOButter42, Freyr, Redvers, GiovanniS, Ttownfeen, Mikenolte, Vital303, Kenyon, Oleg Alexandrov, Mindmatrix, Borb, Rchrd, Commander Keane, Ruud Koot, SergeyLitvinov, Tabletop, Rhun~enwiki, CharlesC, Pingswept, Marudubshinki, RuM, Graham87, Qwertyus, Ryan Norton, Daly, Rjwilmsi, Koavf, Jehochman, Bubba73, Utuado, Yamamoto Ichiro, FlaBot, Arnero, Doc glasgow, Margosbot~enwiki, Chobot, DVdm, Bgwhite, Adoniscik, Peterl, YurikBot, Wavelength, Eirik, Mahahahaneapneap, RussBot, Taejo, Hede2000, Splash, Manop, Gaius Cornelius, Rsrikanth05, Petter Strandmark, Sangwine, Moe Epsilon, LodeRunner, Froth, Misza13, Syrthiss, Xompanthy, Bota47, Sebleblanc, Eli Osherovich, Slicing, EdMiller, Orioane, Closedmouth, Cedar101, Little Savage, Mike1024, JLaTondre, Gesslein, Attilios, Tttrung, SmackBot, Imz, InverseHypercube, Matmota, Pgk, Chaohwa, Bmearns, Mgreenbe, Cool3, Mcld, Gilliam, Benjaminevans82, Skizzik, DocKrin, DStoykov, Thumperward, Oli Filth, MalafayaBot, DHN-bot~enwiki, Tsca.bot, Murder1, Kjetil1001, Berland, Stepho-wrs, Ssnseawolf, EIFY, Shushruth, Mystic Pixel, Nmnogueira, Spiritia, SashatoBot, Shields020, Ojophoyimbo, SS2005, Akbg, RCX, Tim bates, Tktktk, Minna Sora no Shita, Morten, Sebdude, Collect, Hvn0413, Rogerbrent, Dicklyon, Sharcho, Hovden, Norm mit, Melonakos, CapitalR, Tawkerbot2, Lavaka, Ioannes Pragensis, LeRoi, Paul Matthews, Toto76, Raysonho, KyraVixen, Engelec, ZsinjBot, Lehalle, Steveaa, Shoez, Requestion, Shreyasjoshis, Adhanali, Yaris678, Jac16888, Dimacq, FastLizard4, Mostafarazavi, Qwyrxian, JacobBramley, Stanislav87, Tbeu, CynicalMe, Anupam, DmitTrix, West Brom 4ever, Electron9, Peachris, ++Martin++, Dawnseeker2000, Urdutext, Chamilton333, Darklilac, Zchris87v, Softwarehistorian, JAnDbot, GromXXVII, Arifsaha, Vanya, Oxinabox, MER-C, Chaucer1387, Dalek Cab, Magioladitis, Dmoulton, Tomwalters, BeauPaisley, VoABot II, Jarekt, Tedickey, Twsx, Baccyak4H, Cic, SwiftBot, Alex Spade, David Eppstein, User A1, AllenDowney, Gwern, Stephenchou0722, Danman111111, Jtir, MartinBot, Mythealias, Sigmundg, R'n'B, Mange01, Arite, Xcbluedevil, Notreallydavid, BrianOfRugby, Policron, Doc123, KylieTastic, KudzuVine, VolkovBot, ABF, Pleasantville, Nburden, Redgecko, Wikipedante, Cdowley, Amroamroamro, Davehi1, Engginiranjan, Oxfordwang, Inductiveload, Zhangrenpeng93, Itemirus, Doeydoey, Compscigenius, Meters, Stmueller, Bmtran, SiggyF, EmxBot, Anthemion10, Nubiatech, Matthew Yeager, Swaq, Ygramul, Jerryobject, Chiron80, Rmerpes, Flyer22 Reborn, Masgatotkaca, Kgarr, Fttguitarist, Free Software Knight, Oxymoron83, Lourakis, JackSchmidt, Vice regent, Mygerardromance, Randomblue, Baosheng, Tuxa, ClueBot, Ashemon, The Thing That Should Not Be, Pcirrus~enwiki, Boing! said Zebedee, Jamieep, Pmcalduff, Mspraveen, Flaming, DragonBot, Greenmatter, Excirial, Mr squelch, ExterminationXIII, Bender2k14, PixelBot, Eyal.krupka, Muhandes, Stypex, Stdjmax, AmygdalaWiki, MelonBot, Qwfp, DumZiBoT, XLinkBot, T68492, Manishti2004, WikHead, Avish2217, ZooFari, RyanLeiTaiwan, Pioneer42, Addbot, Misterturtie, Fmorstatter, Mortense, Ghettoblaster, Wordsoup, Lowk, Sontofle, Fgnievinski, L.djinevski, Dankdinosaur, SpillingBot, NjardarBot, MrOllie, Favonian, Tinkar, Garykempen, Tide rolls, Lightbot, Zorrobot, Jarble, Luckas-bot, Yobot, Bunnyhop11, Ptbotgourou, SuperSlacker, Jason Recliner, Esq., Mahesh.vibhute, KamikazeBot, AnomieBOT, Rubinbot, Speller26, Yachtsman1, Materialscientist, Lkt1126, Haleyga, Jamawama, ArthurBot, PavelSolin, Premvvnc, Carstorm, Jkbw, Gtfjbl, DanDoherty79, Magnesium, RibotBOT, SassoBot, Mathonius, Leonardo Da Vinci, Aveir, Mcmlxxxi, Stpboyd, Britlak, Captain-n00dle, Sea-monsters, LucienBOT, X7q, Necroplasma, Engineering1, Esjs, Atlantia, ArnaudContet, Winterst, Aniskhan001, Hoo man, EdifChao, Christopher1968, FoxBot, DixonDBot, Francesca.moyse, Mga010, Jesse V., WillNess, Mean as custard, Trabant01, Engmohammedelsaid, Bumblebrie, Mcarone1, J36miles, Siddhartha 90, Kapelson, Gautam73, Mduench, Werieth, John Cline, Cupidvogel, Fæ, ZmeiGorynych, The Dark Melon, 刻意, Ὁ οἶστρος, Laughingwell, Sealbock, Ee.muhammad, Monterey Bay, Lexusuns, TyA, L Kensington, JordiGH, ChuispastonBot, Kunwon.saw, Wukefe, Kaiser czar, Ewharpin, Diftil, DASHBotAV, Petrb, ClueBot NG, EJ257T40R, Satellizer, ImagingGuy, Parcly Taxel, Luca Ghio, Widr, Lazerchikin, Helpful Pixie Bot, Eric.klumpp, Cpltwine, Doorknob747, Bhahas, BG19bot, Bmusician, Sgwfmk8, Flatheman3, Amolbot, Manu31415, Hottiulrich, Snow Blizzard, BigNum, Chanvese, Rtmcrrctr, Bakkedal, Mediran,

YFdyh-bot, F1000F2000, Mllyjn, IjonTichyIjonTichy, Billy Vazquez, Marchuf, Dccarles, Ajinkyaj, Sminthopsis84, Mogism, PeerBaba, Cito-genesis42, Jeetsagar, Isarra (HG), P.arashnia, Reatlas, Krishna indian, Onewordify, User 99 119, AndrewJanke, CleverScience, Rhme, And-faulkner, Ugog Nizdast, Ccmcc2012, ScotXW, Impsswoon, Alienmang, Monkbot, AKS.9955, Buckeyesfan15, TranquilHope, JMP EAX, Pdynes, Tshubham, Atta.ur.rehman.hashmi, Yanga Baracuda, Yolololololo Yolololololo twice, Madarin.eu, NuclearLemon, Gtouchan94, Matlabrob and Anonymous: 660

- **MeVisLab** *Source:* https://en.wikipedia.org/wiki/MeVisLab?oldid=668787167 *Contributors:* Deb, Vadmium, Gadfium, Rubik-wuerfel, Smack-Bot, Mauls, Chris the speller, Ohconfucius, Tedickey, R'n'B, SchreiberBike, Dekart, Yobot, TheCuriousGnome, FrescoBot, DASHBot, Van-ished 1850, Bcgrossmann, Snotbot, BG19bot, BattyBot, Fritter1, Hans Meine, Aleks-ger, Apoplexi and Anonymous: 16

- **Mocolo** *Source:* https://en.wikipedia.org/wiki/Mocolo?oldid=606038093 *Contributors:* Stuartyeates, Michael Slone, Smartyllama, Alaibot, HairyWombat, C6541, Dawynn, Orf Quarenghi, Why should I have a User Name? and Anonymous: 1

- **OpenCV** *Source:* https://en.wikipedia.org/wiki/OpenCV?oldid=682836471 *Contributors:* Komap, AnonMoos, Smb1001, Abdull, Harriv, Vio-letriga, Spoon!, Cyc~enwiki, Ferkel, Mrzaius, Forderud, Oleg Alexandrov, CharlesC, Qwertyus, JIP, Husky, W3bbo, FlaBot, Chninkel, Tedder, Chobot, Pegship, Open2universe, Femmina, JLaTondre, Garybradski, SmackBot, Lukas Mach, KYN, Skizzik, Thumperward, Frap, Olddocks, Khaled.khalil, Guroadrunner, Samfreed, Aboeing, Hu12, Arto B, Smomen, CmdrObot, Jokes Free4Me, Vladimir Drzik, Waltgibson, Mailseth, Gioto, Filnik, Tedickey, Correll, EdBever, Jiuguang Wang, DeeKay64, Usp, Teacurran, VolkovBot, Rei-bot, Shiqi Yu~enwiki, McM.bot, Per-ryTachett, Aleks-eng, Brethren psp, Samontab, XLinkBot, RyanLeiTaiwan, Addbot, Pyerre, Mortense, Carcediano, AndrewHZ, Shervine-mami, MrOllie, Ollydbg, Zorrobot, Yobot, Canming, Flavio58, D0ktorz, AnomieBOT, Grbradsk, Rjanag, Peval27, ArthurBot, Padrem, VCSBC64xx, Martnym, Dexmac, 777sms, Alireza.haghshenas, Migaber, Jdbtwo, EmausBot, Timscaffidi, Erget2005, Blibrestez55, Mikhail Ryazanov, Helpful Pixie Bot, Bsd noobz, DBigXray, Hurricanefan25, Happyuk, Vswitchs, Rmgash, ChrisGualtieri, Mkwilliams87, Ploki-jnu, Sgarizvi1991, Sai123k, Gary Bradski, Lone boatman, MaatyBot, Freofalls, Babitaarora, Djibi2, Aleks-ger, Samratsubedi, Patrios, Ab-hishek4273, Rober9876543210, EfiHerbst and Anonymous: 124

- **Pfinder** *Source:* https://en.wikipedia.org/wiki/Pfinder?oldid=617908713 *Contributors:* Bearcat, Rjwilmsi, CmdrObot, GordiasAchos, Katharineamy, Matt c clark and Monkbot

- **Pipeline Pilot** *Source:* https://en.wikipedia.org/wiki/Pipeline_Pilot?oldid=685486154 *Contributors:* Darkwind, Inkling, Daniel Bonniot de Ruisselet, Kkmurray, Cydebot, Krushia, Stooof, Yobot, Citation bot, BattyBot, Rybec and Anonymous: 1

- **RapidMiner** *Source:* https://en.wikipedia.org/wiki/RapidMiner?oldid=689861158 *Contributors:* Enchanter, Kwertii, Kku, Ronz, Den fjät-trade ankan~enwiki, Darkwind, Jm34harvey, SEWilco, Khalid hassani, Golbez, Utcursch, Pgan002, Urhixidur, Andreas Kaufmann, Thor-wald, CanisRufus, John Vandenberg, Alexander Sommer, Diego Moya, Karnesky, GregorB, FlaBot, Mfranck, NeilN, Rwwww, SmackBot, Bluebot, JonHarder, MatthewKarlsen, Krexer, Derek farn, Ralf Klinkenberg, Ioannes Pragensis, Nczempin, Mirka~enwiki, Cydebot, Hum-bleGod, Northumbrian, Gioto, Vanya, Ingomierswa, Wonton, R'n'B, Tgeairn, KylieTastic, Nabil2199, Djmckee1, Eagleal, Free Software Knight, ImageRemovalBot, Beeblebrox, Ottawahitech, Chaosdruid, 1ForTheMoney, Nikhilkrgvr, XLinkBot, Addbot, Ronhjones, MrOllie, SpBot, Lightbot, Luckas-bot, Yobot, LeonardoWeiss, AnomieBOT, ThaddeusB, Alphmega100, Alvin Seville, Alainr345, LittleWink, Ace man35, Dewritech, Sgoder, Chire, Palosirkka, Doctorambient, Luke145, PhnomPencil, NikitaPSolanki, Pierowski, Melchiaros, Soulparadox, Terada, Editfromwithout, Lemnaminor, Pdecalculus, Comp.arch, Goattender, Shavaiz Shams, Filedelinkerbot, Rober9876543210, Clbsierra, Messecity and Anonymous: 56

- **RoboRealm** *Source:* https://en.wikipedia.org/wiki/RoboRealm?oldid=615624443 *Contributors:* Nealmcb, Macl, Dialectric, SmackBot, KYN, NatGertler, Katharineamy, Meters, CorenSearchBot, Sun Creator, AstarothN, Yobot, AnomieBOT, A little insignificant, ExDxV, Danim, Scopecreep, Steven.Gentner and Anonymous: 7

- **Robot Operating System** *Source:* https://en.wikipedia.org/wiki/Robot_Operating_System?oldid=688654273 *Contributors:* Berek, Daniel Mahu, Ezra Wax, Baylink, BAxelrod, Sabbut, Poo-T~enwiki, Pawpawyoung, YUL89YYZ, Stephane.magnenat, Scott Ritchie, Voxadam, Oleg Alexandrov, Tony1, JLaTondre, Rwwww, Matt Heard, SmackBot, KYN, Chris the speller, Frap, Charles Merriam, Rolfedh, Gioto, Guy Macon, Coolg49964, Trasz, Rocketmagnet, Softtest123, Jerryobject, JL-Bot, Robotilak, Niceguyedc, Arjayay, Kolyma, MystBot, Ad-dbot, Mortense, Abovechief, Rasulnrasul, Luckas-bot, Yobot, Grbradsk, Tbfnews, Xqbot, FrescoBot, W130s, HRoestBot, Jbohren, EmausBot, Zollerriia, ZéroBot, Gordon1104, Antefuturisme, Luis Mailhos, Danim, RobertPollak, Feuerhand, BG19bot, Mschuldt, BattyBot, Autodidak-tos, Comatmebro, Alb d91, Alismayilov, Robocero, Ra ules, Al9010, Odestcj, Hari Seldon, Tim36272, Lagoset, Esteve.fernandez, Rahugur, Velvel2, Aspotlessmind, Paul Hvass and Anonymous: 41

- **Scilab Image Processing** *Source:* https://en.wikipedia.org/wiki/Scilab_Image_Processing?oldid=547342748 *Contributors:* KYN, Slakr, Ad-hanali, Bobblehead, Gioto, Liberio, ClueBot, Addbot, PlankBot, Ulric1313, Bomazi, MerllwBot, Liberio1998 and Anonymous: 8

- **SigmaScan** *Source:* https://en.wikipedia.org/wiki/SigmaScan?oldid=660103641 *Contributors:* Rubik-wuerfel, BD2412, Rjwilmsi, Treygdor, EdGl, Alaibot, BetacommandBot, DGG, Carmen56, SPMarketing, Skinnyg5, 3gg5ample and Anonymous: 3

- **Softwarp** *Source:* https://en.wikipedia.org/wiki/Softwarp?oldid=653429075 *Contributors:* Nick Number, SoxBot, Addbot, Yobot, Fraggle81, DemocraticLuntz, Robban75, Erik9bot, AvicAWB, DoctorKubla, Webclient101, ImmersaView, Christine YM Zhang, Adriatik neziri, Vel-murugann and Anonymous: 6

- **Trax Image Recognition** *Source:* https://en.wikipedia.org/wiki/Trax_Image_Recognition?oldid=680317998 *Contributors:* Niceguyedc, OluwaCur-tis, Computervision15 and Anonymous: 1

- **VDSI** *Source:* https://en.wikipedia.org/wiki/VDSI?oldid=668893828 *Contributors:* Ceyockey, Woohookitty, KYN, GoodDay, Alaibot, Mbr4, Addbot, Dawynn, Mo ainm, Fred.Pendleton and Anonymous: 3

- **VIGRA** *Source:* https://en.wikipedia.org/wiki/VIGRA?oldid=607450437 *Contributors:* KYN, Dekart, Chire and Qingzhong

- **VIRAT** *Source:* https://en.wikipedia.org/wiki/VIRAT?oldid=646911253 *Contributors:* Vsion, Jrtayloriv, King of Hearts, Wavelength, Smack-Bot, Frap, Pwjb, H3llBot, BattyBot, Michipedian, Itstabishaaftab and Anonymous: 3

- **VTK** *Source:* https://en.wikipedia.org/wiki/VTK?oldid=684229907 *Contributors:* Kku, RichiH, Pengo, Pascal666, Delta G, Abdull, Zyqqh, Marasmusine, Waldir, Rufous, Qwertyus, Pdelong, FlaBot, CambridgeBayWeather, Spike Wilbury, Ooble, Ojii-san, JLaTondre, Carlos Capdepón, SmackBot, KYN, Fuzzform, Frap, Fuhghettaboutit, Daniel.Cardenas, Digana, LuisIbanez, Lorensen, Pit-yacker, Cydebot, AntiVandalBot, NE2, Richard Giuly, Gwern, R'n'B, Dzenanz, Synthebot, CMBJ, Free Software Knight, Sfan00 IMG, Kl4m-AWB, Auntof6, Htnahsarp, SchreiberBike, Botalex, PretentiousSnot, CallipygianSchoolGirl, Zeliboba7, Addbot, Mortense, Tothwolf, Bnwylie, Zezao, Yobot, FrescoBot, LucienBOT, Pratik.mallya, Rodamaker, Foobarnix, Tim1357, RMartin-2, KWComm, Alisha.4m, Logic cube, Lorem Ip, Karima Rafes, ChuispastonBot, Bcgrossmann, Helpful Pixie Bot, Sergtk, MrBrunchtime, Filiagees, Julia Abril, AID4wiki, ScotXW, EfiHerbst, Nc4096 and Anonymous: 39

- **VXL** *Source:* https://en.wikipedia.org/wiki/VXL?oldid=640896089 *Contributors:* Alai, Gussisaurio, Nachoman-au, Timchanzee, KYN, ST47, Daviddoria, Dawynn, Jonas AGX, Bineeshmail, Jsayre64, ChrisGualtieri and Anonymous: 5

- **YaDICs** *Source:* https://en.wikipedia.org/wiki/YaDICs?oldid=681863739 *Contributors:* JHCaufield, Mild Bill Hiccup, AnomieBOT, BG19bot, Aus0107, Lakun.patra, OMPIRE, JeffWitz and Anonymous: 1

44.5.2 Images

- **File:3DSlicerLogo.png** *Source:* https://upload.wikimedia.org/wikipedia/en/b/be/3DSlicerLogo.png *License:* Fair use *Contributors:* The logo is from the slicer.org website. slicer.org
  Original artist: ?

- **File:3dReconstruct_Mechanical_Part.png** *Source:* https://upload.wikimedia.org/wikipedia/commons/8/85/3dReconstruct_Mechanical_Part.png *License:* CC BY-SA 3.0 *Contributors:* Visualization Sciences Group (VSG) *Original artist:* Egallois (talk)

- **File:Aforgenet.jpg** *Source:* https://upload.wikimedia.org/wikipedia/commons/b/bb/Aforgenet.jpg *License:* Public domain *Contributors:* http://www.aforgenet.com/img/aforgenetf.jpg *Original artist:* Andrew Kirillov

- **File:Ambox_important.svg** *Source:* https://upload.wikimedia.org/wikipedia/commons/b/b4/Ambox_important.svg *License:* Public domain *Contributors:* Own work, based off of Image:Ambox scales.svg *Original artist:* Dsmurat (talk · contribs)

- **File:Amira_Screenshot_with_Honeybee_Brain_visualization.png** *Source:* https://upload.wikimedia.org/wikipedia/commons/0/05/Amira_Screenshot_with_Honeybee_Brain_visualization.png *License:* CC BY-SA 3.0 *Contributors:* Own work *Original artist:* Christian Wietholt

- **File:Analysisdata.jpg** *Source:* https://upload.wikimedia.org/wikipedia/en/5/5f/Analysisdata.jpg *License:* Cc-by-sa-3.0 *Contributors:* Own work
  Original artist:
  Way4thesub (talk) (Uploads)

- **File:Android_robot.svg** *Source:* https://upload.wikimedia.org/wikipedia/commons/d/d7/Android_robot.svg *License:* CC BY 3.0 *Contributors:* Sítio com material do Android (ficheiros GIF e PS): *Original artist:* Google Inc.

- **File:Animation2.gif** *Source:* https://upload.wikimedia.org/wikipedia/commons/c/c0/Animation2.gif *License:* CC-BY-SA-3.0 *Contributors:* Own work *Original artist:* MG (talk · contribs)

- **File:Avizo-earth_GUI.jpg** *Source:* https://upload.wikimedia.org/wikipedia/en/8/80/Avizo-earth_GUI.jpg *License:* CC-BY-SA-3.0 *Contributors:*
  Visualization Sciences Group (VSG)
  Original artist:
  Egallois (talk)

- **File:Avizo-fire_GUI.png** *Source:* https://upload.wikimedia.org/wikipedia/en/c/c9/Avizo-fire_GUI.png *License:* CC-BY-SA-3.0 *Contributors:*
  Visualization Sciences Group (VSG)
  Original artist:
  Egallois (talk)

- **File:Avizo_3D_imaging_and_analysis_software_logo.jpg** *Source:* https://upload.wikimedia.org/wikipedia/commons/c/c1/Avizo_3D_imaging_and_analysis_software_logo.jpg *License:* CC BY-SA 3.0 *Contributors:* Own work *Original artist:* Egallois

- **File:Ayotle_logo.png** *Source:* https://upload.wikimedia.org/wikipedia/commons/2/2b/Ayotle_logo.png *License:* CC BY-SA 3.0 *Contributors:* Own work *Original artist:* KyungJuneK

- **File:Ballerina-icon.jpg** *Source:* https://upload.wikimedia.org/wikipedia/commons/3/3a/Ballerina-icon.jpg *License:* CC-BY-SA-3.0 *Contributors:*

- Snowdance.jpg *Original artist:* Snowdance.jpg: Rick Dikeman

- **File:Bing_logo_(2013).svg** *Source:* https://upload.wikimedia.org/wikipedia/commons/b/b1/Bing_logo_%282013%29.svg *License:* Public domain *Contributors:* The logo comes from bing.com/explore/newbing. SVG version is created by AxG (talk · contribs). *Original artist:* Logo is made by Microsoft. SVG version is created and uploaded by AxG (talk · contribs).

- **File:CVIPtoolsfrontpage.png** *Source:* https://upload.wikimedia.org/wikipedia/commons/5/54/CVIPtoolsfrontpage.png *License:* Copyrighted free use *Contributors:* Own work *Original artist:* Samratsubedi

- **File:CancerExample.jpg** *Source:* https://upload.wikimedia.org/wikipedia/en/2/25/CancerExample.jpg *License:* Cc-by-sa-3.0 *Contributors:* Own work
  Original artist:
  Way4thesub (talk) (Uploads)

- **File:Cart_pushing_rviz_holonomic.jpg** *Source:* https://upload.wikimedia.org/wikipedia/commons/b/b8/Cart_pushing_rviz_holonomic.jpg *License:* CC BY 3.0 *Contributors:* http://www.ros.org/wp-content/uploads/2013/12/cart_pushing_rviz_holonomic.jpg *Original artist:* Open Source Robotics Foundation

- **File:Commons-logo.svg** *Source:* https://upload.wikimedia.org/wikipedia/en/4/4a/Commons-logo.svg *License:* ? *Contributors:* ? *Original artist:* ?

- **File:Crystal_Clear_device_cdrom_unmount.png** *Source:* https://upload.wikimedia.org/wikipedia/commons/1/10/Crystal_Clear_device_cdrom_unmount.png *License:* LGPL *Contributors:* All Crystal Clear icons were posted by the author as LGPL on kde-look; *Original artist:* Everaldo Coelho and YellowIcon;

- **File:Deep_Dream_Toast_Sandwich.jpg** *Source:* https://upload.wikimedia.org/wikipedia/en/4/47/Deep_Dream_Toast_Sandwich.jpg *License:* CC-BY-SA-3.0 *Contributors:*
  This image was made available under the GNU Free Documentation License.
  Previously published: Published here first. *Original artist:*
  DoctorWho42

- **File:Dlib_c++_library_logo.png** *Source:* https://upload.wikimedia.org/wikipedia/en/d/d9/Dlib_c%2B%2B_library_logo.png *License:* Fair use *Contributors:* From dlib mercurial repository containing the vector graphics logo file. Repository is located at http://hg.code.sf.net/p/dclib/code *Original artist:* ?

- **File:Emoji_u1f4f1.svg** *Source:* https://upload.wikimedia.org/wikipedia/commons/1/17/Emoji_u1f4f1.svg *License:* Apache License 2.0 *Contributors:* https://code.google.com/p/noto/ *Original artist:* Google

- **File:FIJI_(software)_Logo.svg** *Source:* https://upload.wikimedia.org/wikipedia/commons/5/55/FIJI_%28software%29_Logo.svg *License:* Public domain *Contributors:* Own work *Original artist:* Benjamin Pavie but originaly designed by the FIJI team

- **File:Fibertracking_wp_B.jpg** *Source:* https://upload.wikimedia.org/wikipedia/en/c/c6/Fibertracking_wp_B.jpg *License:* CC-BY-SA-3.0 *Contributors:*
  http://www.mevislab.de/mevislab/screenshots/ *Original artist:*
  Fraunhofer MEVIS

- **File:Folder_Hexagonal_Icon.svg** *Source:* https://upload.wikimedia.org/wikipedia/en/4/48/Folder_Hexagonal_Icon.svg *License:* Cc-by-sa-3.0 *Contributors:* ? *Original artist:* ?

- **File:Free_Software_Portal_Logo.svg** *Source:* https://upload.wikimedia.org/wikipedia/commons/6/67/Nuvola_apps_emacs_vector.svg *License:* LGPL *Contributors:*

- Nuvola_apps_emacs.png *Original artist:* Nuvola_apps_emacs.png: David Vignoni

- **File:GemIdentComposite.jpg** *Source:* https://upload.wikimedia.org/wikipedia/commons/6/65/GemIdentComposite.jpg *License:* CC-BY-SA-3.0 *Contributors:* Transferred from en.wikipedia by Ronhjones *Original artist:* Way4thesub at en.wikipedia

- **File:GemIdent_logo.jpg** *Source:* https://upload.wikimedia.org/wikipedia/en/d/d6/GemIdent_logo.jpg *License:* Fair use *Contributors:*
  The logo may be obtained from GemIdent.
  Original artist: ?

- **File:GimiasLogo.jpg** *Source:* https://upload.wikimedia.org/wikipedia/commons/4/44/GimiasLogo.jpg *License:* Public domain *Contributors:* GIMIAS website *Original artist:* cistib

- **File:Ginkgo_CADx_displaying_a_CT.jpg** *Source:* https://upload.wikimedia.org/wikipedia/en/a/a1/Ginkgo_CADx_displaying_a_CT.jpg *License:* CC-BY-SA-3.0 *Contributors:*
  Making an screenshot.
  Previously published: This is an original work. It isn't published anywhere.
  Original artist:
  Metadiego

- **File:Google_Goggles.png** *Source:* https://upload.wikimedia.org/wikipedia/en/1/1a/Google_Goggles.png *License:* Fair use *Contributors:* http://www.enterakt.com/wp-content/uploads/2011/01/GoogleGoggles.jpg *Original artist:* ?

- **File:Gvr_body_wp.jpg** *Source:* https://upload.wikimedia.org/wikipedia/en/b/b0/Gvr_body_wp.jpg *License:* CC-BY-SA-3.0 *Contributors:*
  http://www.mevislab.de/mevislab/screenshots/ *Original artist:*
  Fraunhofer MEVIS

- **File:Gvrheart2_wp.jpg** *Source:* https://upload.wikimedia.org/wikipedia/en/4/4c/Gvrheart2_wp.jpg *License:* CC-BY-SA-3.0 *Contributors:*
  http://www.mevislab.de/mevislab/screenshots/ *Original artist:*
  Fraunhofer MEVIS

- **File:Gvrheart_wp.jpg** *Source:* https://upload.wikimedia.org/wikipedia/en/1/18/Gvrheart_wp.jpg *License:* CC-BY-SA-3.0 *Contributors:*
  http://www.mevislab.de/mevislab/screenshots/ *Original artist:*
  Fraunhofer MEVIS

- **File:Gvrimage4_05_wp.jpg** *Source:* https://upload.wikimedia.org/wikipedia/en/a/ac/Gvrimage4_05_wp.jpg *License:* CC-BY-SA-3.0 *Contributors:*
  http://www.mevislab.de/mevislab/screenshots/ *Original artist:*
  Fraunhofer MEVIS

- **File:Gvrimage_2_04_wp.jpg** *Source:* https://upload.wikimedia.org/wikipedia/en/1/1f/Gvrimage_2_04_wp.jpg *License:* CC-BY-SA-3.0 *Contributors:*
  http://www.mevislab.de/mevislab/screenshots/ *Original artist:*
  Fraunhofer MEVIS

- **File:ILNumerics_Array_Visualizer_in_a_Visual_Studio_debug_session.png** *Source:* https://upload.wikimedia.org/wikipedia/commons/
  b/b5/ILNumerics_Array_Visualizer_in_a_Visual_Studio_debug_session.png *License:* CC BY-SA 4.0 *Contributors:* Own work *Original artist:*
  Numbers303

- **File:ILPanel_graphical_plots_in_a_Windows_application.png** *Source:* https://upload.wikimedia.org/wikipedia/commons/e/ee/ILPanel_
  graphical_plots_in_a_Windows_application.png *License:* CC BY-SA 4.0 *Contributors:* Own work *Original artist:* Numbers303

- **File:ImageNet_Designer.png** *Source:* https://upload.wikimedia.org/wikipedia/commons/f/f0/ImageNet_Designer.png *License:* CC BY-SA
  3.0 *Contributors:* Screenshot *Original artist:* Uwelange82

- **File:Indigoigloo_600.png** *Source:* https://upload.wikimedia.org/wikipedia/commons/9/91/Indigoigloo_600.png *License:* CC BY 3.0 *Contributors:* http://wiki.ros.org/indigo *Original artist:* http://wiki.ros.org

- **File:InsightToolkitLogo.jpg** *Source:* https://upload.wikimedia.org/wikipedia/commons/c/cd/InsightToolkitLogo.jpg *License:* CC BY 2.5
  Contributors: http://en.wikipedia.org/wiki/File:InsightToolkitLogo.jpg *Original artist:* Logo designed by Julien Jomier for the Insight Software
  Consortium

- **File:MATLAB_R2013a_Win8_screenshot.png** *Source:* https://upload.wikimedia.org/wikipedia/en/e/e1/MATLAB_R2013a_Win8_screenshot.
  png *License:* Fair use *Contributors:*
  self-made screenshot (2013-08-14)
  Original artist: ?

- **File:MATLAB_mesh_sinc3D.svg** *Source:* https://upload.wikimedia.org/wikipedia/commons/c/c4/MATLAB_mesh_sinc3D.svg *License:* Public domain *Contributors:* Own work *Original artist:* DmitTrix

- **File:MATLAB_surf_sinc3D.svg** *Source:* https://upload.wikimedia.org/wikipedia/commons/1/16/MATLAB_surf_sinc3D.svg *License:* Public domain *Contributors:* Own work *Original artist:* DmitTrix

- **File:M_box.svg** *Source:* https://upload.wikimedia.org/wikipedia/commons/9/94/M_box.svg *License:* Public domain *Contributors:* Own work
  based on: File:Microsoft.svg *Original artist:* Ariesk47 (<a href='//commons.wikimedia.org/wiki/User_talk:Ariesk47' title='User talk:Ariesk47'>talk</a>)

- **File:Matlab3.5eastereggs.TIF** *Source:* https://upload.wikimedia.org/wikipedia/commons/b/be/Matlab3.5eastereggs.TIF *License:* CC BY-
  SA 3.0 *Contributors:* Own work *Original artist:* Alienmang

- **File:Matlab_Logo.png** *Source:* https://upload.wikimedia.org/wikipedia/commons/2/21/Matlab_Logo.png *License:* Public domain *Contributors:* Own work *Original artist:* Jarekt

- **File:Matlab_plot_sin.svg** *Source:* https://upload.wikimedia.org/wikipedia/commons/1/1c/Matlab_plot_sin.svg *License:* CC BY-SA 2.5 *Contributors:* uploaded to the English Wikipedia in November 2006 (File:Matlab_plot_sin.svg log) *Original artist:* Nuno Nogueira (talk)

- **File:Merge-arrow.svg** *Source:* https://upload.wikimedia.org/wikipedia/commons/a/aa/Merge-arrow.svg *License:* Public domain *Contributors:* ? *Original artist:* ?

- **File:Mevislab_logo.png** *Source:* https://upload.wikimedia.org/wikipedia/en/5/5a/Mevislab_logo.png *License:* Fair use *Contributors:*
  MeVisLab documentation, available with the free installer
  Original artist: ?

- **File:Mevislab_macosx_wp_B.jpg** *Source:* https://upload.wikimedia.org/wikipedia/en/e/e0/Mevislab_macosx_wp_B.jpg *License:* CC-BY-
  SA-3.0 *Contributors:*
  http://www.mevislab.de/mevislab/screenshots/ *Original artist:*
  Fraunhofer MEVIS

- **File:MocoloUISample.png** *Source:* https://upload.wikimedia.org/wikipedia/en/f/f4/MocoloUISample.png *License:* PD *Contributors:* ? *Original artist:* ?

- **File:OfxOpenCV.png** *Source:* https://upload.wikimedia.org/wikipedia/commons/8/87/OfxOpenCV.png *License:* MIT *Contributors:* Screenshot of openFrameworks *Original artist:* ?

- **File:OpenCV_Logo_with_text_svg_version.svg** *Source:* https://upload.wikimedia.org/wikipedia/commons/3/32/OpenCV_Logo_with_text_
  svg_version.svg *License:* Public domain *Contributors:* http://opencv.willowgarage.com/wiki/OpenCVLogo *Original artist:* Adi Shavit

- **File:Optimization_application_finding_the_minimium_of_the_camel_3_function.png** *Source:* https://upload.wikimedia.org/wikipedia/
  commons/b/b4/Optimization_application_finding_the_minimium_of_the_camel_3_function.png *License:* CC BY-SA 4.0 *Contributors:* Own
  work *Original artist:* Numbers303

- **File:OrangeExample.jpg** *Source:* https://upload.wikimedia.org/wikipedia/en/f/f1/OrangeExample.jpg *License:* Cc-by-sa-3.0 *Contributors:*
  ? *Original artist:* ?

- **File:People_icon.svg** *Source:* https://upload.wikimedia.org/wikipedia/commons/3/37/People_icon.svg *License:* CC0 *Contributors:* OpenClipart *Original artist:* OpenClipart

- **File:Portal-puzzle.svg** *Source:* https://upload.wikimedia.org/wikipedia/en/f/fd/Portal-puzzle.svg *License:* Public domain *Contributors:* ? *Original artist:* ?

- **File:Question_book-new.svg** *Source:* https://upload.wikimedia.org/wikipedia/en/9/99/Question_book-new.svg *License:* Cc-by-sa-3.0 *Contributors:*
  Created from scratch in Adobe Illustrator. Based on Image:Question book.png created by User:Equazcion *Original artist:*
  Tkgd2007

- **File:Science-symbol-2.svg** *Source:* https://upload.wikimedia.org/wikipedia/commons/7/75/Science-symbol-2.svg *License:* CC BY 3.0 *Contributors:* en:Image:Science-symbol2.png *Original artist:* en:User:AllyUnion, User:Stannered

- **File:Software_spanner.png** *Source:* https://upload.wikimedia.org/wikipedia/commons/8/82/Software_spanner.png *License:* CC-BY-SA-3.0 *Contributors:* Transferred from en.wikipedia; transfer was stated to be made by User:Rockfang. *Original artist:* Original uploader was CharlesC at en.wikipedia

- **File:Symbol_list_class.svg** *Source:* https://upload.wikimedia.org/wikipedia/en/d/db/Symbol_list_class.svg *License:* Public domain *Contributors:* ? *Original artist:* ?

- **File:Synaptic.png** *Source:* https://upload.wikimedia.org/wikipedia/commons/0/05/Synaptic.png *License:* GPL *Contributors:* [1] *Original artist:* en:User:Burgundavia

- **File:Unbalanced_scales.svg** *Source:* https://upload.wikimedia.org/wikipedia/commons/f/fe/Unbalanced_scales.svg *License:* Public domain *Contributors:* ? *Original artist:* ?

- **File:VIRAT--_system_concept_diagram.jpg** *Source:* https://upload.wikimedia.org/wikipedia/commons/5/54/VIRAT--_system_concept_diagram.jpg *License:* Public domain *Contributors:* http://www.darpa.mil/ipto/solicit/baa/BAA-08-20.pdf *Original artist:* DARPA / Information Processing Technology Office

- **File:VIRAT-operational-concept-diagram.jpg** *Source:* https://upload.wikimedia.org/wikipedia/commons/a/af/VIRAT-operational-concept-diagram.jpg *License:* Public domain *Contributors:* http://www.darpa.mil/ipto/solicit/baa/BAA-08-20.pdf *Original artist:* Wikipedia: DARPA / Wikipedia: Information Processing Technology Office

- **File:VTKlogo.png** *Source:* https://upload.wikimedia.org/wikipedia/en/e/eb/VTKlogo.png *License:* Fair use *Contributors:*
  The logo may be obtained from VTK.
  Original artist: ?

- **File:Vessels_wp.jpg** *Source:* https://upload.wikimedia.org/wikipedia/en/c/c5/Vessels_wp.jpg *License:* CC-BY-SA-3.0 *Contributors:*
  http://www.mevislab.de/mevislab/screenshots/ *Original artist:*
  Fraunhofer MEVIS

- **File:VirtualAngiography2.jpg** *Source:* https://upload.wikimedia.org/wikipedia/commons/c/c8/VirtualAngiography2.jpg *License:* BSD *Contributors:* software screenshot *Original artist:* ?

- **File:Virtual_permeameter_for_absolute_permeability_computation.jpeg** *Source:* https://upload.wikimedia.org/wikipedia/en/2/29/Virtual_permeameter_for_absolute_permeability_computation.jpeg *License:* CC-BY-3.0 *Contributors:* ? *Original artist:* ?

- **File:Vista-arts.png** *Source:* https://upload.wikimedia.org/wikipedia/commons/4/4e/Vista-arts.png *License:* GPL *Contributors:* ? *Original artist:* ?

- **File:Visual_Studio_code_debuggin_with_ILNumerics_Array_Visualizer.png** *Source:* https://upload.wikimedia.org/wikipedia/commons/9/91/Visual_Studio_code_debuggin_with_ILNumerics_Array_Visualizer.png *License:* CC BY-SA 4.0 *Contributors:* Own work *Original artist:* Numbers303

- **File:Wikibooks-logo-en-noslogan.svg** *Source:* https://upload.wikimedia.org/wikipedia/commons/d/df/Wikibooks-logo-en-noslogan.svg *License:* CC BY-SA 3.0 *Contributors:* Own work *Original artist:* User:Bastique, User:Ramac et al.

- **File:Wikiversity-logo.svg** *Source:* https://upload.wikimedia.org/wikipedia/commons/9/91/Wikiversity-logo.svg *License:* CC BY-SA 3.0 *Contributors:* Snorky (optimized and cleaned up by verdy_p) *Original artist:* Snorky (optimized and cleaned up by verdy_p)

- **File:Windows_logo_-_2012_(red).svg** *Source:* https://upload.wikimedia.org/wikipedia/commons/f/f8/Windows_logo_-_2012_%28red%29.svg *License:* Public domain *Contributors:* This file is derived from File:Windows logo - 2012.svg. *Original artist:* Microsoft Corporation

- **File:Zune_logo_and_wordmark.svg** *Source:* https://upload.wikimedia.org/wikipedia/commons/8/82/Zune_logo_and_wordmark.svg *License:* Public domain *Contributors:* **Original work:** zune.net (direct link)
  Original artist: **This work:** Svgalbertian of English Wikipedia

44.5.3   Content license

- Creative Commons Attribution-Share Alike 3.0

www.ingramcontent.com/pod-product-compliance
Lightning Source LLC
Chambersburg PA
CBHW080810180526
45168CB00006B/2397

* 9 7 8 1 5 1 9 2 2 0 2 3 3 *